SUSTAINABLE AVIATION FUTURES

TRANSPORT AND SUSTAINABILITY

Series Editors: Stephen Ison and Jon Shaw

Recent Volumes:

TRANSPORT AND SUSTAINABILITY VOLUME 4

SUSTAINABLE AVIATION FUTURES

EDITED BY

LUCY BUDD
Loughborough University, Loughborough, UK

STEVEN GRIGGS
De Montfort University, Leicester, UK

DAVID HOWARTH
University of Essex, Colchester, UK

E·S·R·C ECONOMIC & SOCIAL RESEARCH COUNCIL

Emerald

United Kingdom – North America – Japan
India – Malaysia – China

Emerald Group Publishing Limited
Howard House, Wagon Lane, Bingley BD16 1WA, UK

First edition 2013

British Library Cataloguing in Publication Data
A catalogue record for this book is available from the British Library

ISBN: 978-1-78190-595-1
ISSN: 2044-9941 (Series)

ISOQAR
REGISTERED
Certificate Number 1985
ISO 14001

ISOQAR certified Management System, awarded to Emerald for adherence to Environmental standard ISO 14001:2004.

INVESTOR IN PEOPLE

CONTENTS

LIST OF CONTRIBUTORS

Peter Adey	Department of Geography, Royal Holloway, University of London, Egham, Surrey, UK
Kevin Anderson	School of Mechanical, Civil and Aerospace Engineering and the Tyndall Centre for Climate Change Research, University of Manchester, Manchester, UK
Annela Anger-Kraavi	School of Environmental Sciences University of East Anglia, Norwich Research Park, Norwich, UK
John Bowen	Department of Geography, Central Washington University, Ellensburg, WA, USA
Alice Bows-Larkin	School of Mechanical, Aerospace and Civil Engineering, Tyndall Centre for Climate Change Research, University of Manchester, Manchester, UK
Christian Bröer	University of Amsterdam, Department of Sociology and Cultural Anthropology, Amsterdam, The Netherlands
Lucy Budd	Transport Studies Group, School of Civil and Building Engineering, Loughborough University, Leicestershire, UK
Thomas Budd	Department of Geography & Environment, University of Aberdeen, Aberdeen, UK
James Connelly	School of Politics, Philosophy and International Studies, University of Hull, Hull, UK

Frédéric Dobruszkes Transport Studies Unit, University of Oxford, Oxford, UK

Moshe Givoni Department of Geography and the Human Environment, Tel Aviv University, Tel Aviv, Israel

Steven Griggs Leicester Business School, The Gateway, De Montfort University, Leicester, UK

Charlotte Halpern Centre d'Etudes Européennes de Sciences Po, Paris, France

David Howarth Department of Government, University of Essex, Colchester, Essex, UK

Ute Knippenberger Head of Urban Planning, City of KronbergIm Taunus, Frankfurt, Germany

Jonathan Köhler Fraunhofer-Institut für System- und Innovationsforschung ISI, Karlsruhe, Germany

PART I
CONTEXT

CHAPTER 1

SUSTAINABLE AVIATION FUTURES: CRISES, CONTESTED REALITIES AND PROSPECTS FOR CHANGE

Lucy Budd, Steven Griggs and David Howarth

ABSTRACT

Purpose – *This chapter examines the torsions and blind spots that structure the contemporary debate on the politics and policy of aviation. It also generates different scenarios for the future of air travel, which can help to unblock the current impasse about the perceived costs and benefits of aviation and its attendant infrastructural needs.*

Originality – *This chapter characterises and evaluates the competing frames that organise the contested realities of air transport. By mapping out the current fault lines of aviation politics and policy, the chapter is also able to delineate four main scenarios regarding the future of aviation, which we name the 'post-carbon', 'high-modernist', 'market regulation' and 'demand management' projections respectively.*

Methodology/approach – *The chapter problematises and criticises the existing literature, policy reports and stakeholder briefings that inform*

Sustainable Aviation Futures
Transport and Sustainability, Volume 4, 3–35

ISSN: 2044-9941/doi:10.1108/S2044-9941(2013)0000004013

the contemporary standoff in UK aviation policy. It uses the definition of sustainable development as a heuristic device to map and identify the fault lines structuring contemporary debates on aviation futures. It then builds upon this analysis to delimit four different scenarios for the future of flying.

Findings – *The chapter analyses the contested realities of aviation politics. It re-affirms the political nature of such divisions, which in turn structure the rival understandings of aviation. The analysis suggests that the identified fault lines are constantly reiterated by competing appeals to ambiguous and contradictory evidence-bases or policy frames. Ultimately, the chapter claims that any significant reframing of aviation policy and politics rests on the outcome of political negotiations and persuasion. But it also depends on the broader views of citizens and stakeholders about the future challenges facing society, as well as the way in which governments and affected agents put in place and coordinate the multiple arenas in which a dialogue over the future of aviation can be held. Aviation futures cannot be reduced to the narrow confines of the technical merits or claims surrounding the feasibility of policy instruments.*

Keywords: Air travel; airports; aviation economics; sustainable development; environment; aviation futures

'Crisis' is a term often swiftly attached to industries, governments and indeed societies, only to be withdrawn at some later date as the expected threats fail to appear or are met with new policy instruments and adjustments. Such crises are, in fact, relatively easily constrained, capable of being offset by policy learning or changes to our standard ways of working. Yet, recourse to such crisis-narratives should not always be dismissed as 'crying wolf'. The threats and dangers are often very material indeed, to the point where some cannot be simply warded off by innovative or ingenious strategies. Rather, they require the generation of alternative futures, which recast the existing regimes, systems, structures and practices through which an industry, government or society operates. Put alternatively, 'crisis' may well in certain conjunctures demand the radical transformation of how we view the world, putting into question our fundamental values and beliefs; and for this to happen, protagonists have arguably to recognise that they face a 'crisis' and that continuing as usual is no longer an option.

Whether the commercial aviation industry faces a crisis is a moot point, generating different judgements and controversies which in turn quickly trigger further rounds of questions and debates. Not least among such questions is that if aviation is in crisis, it is a crisis of what and for whom? For some commentators, and not just those advancing future imaginaries of mega-airport cities or aerotropoli (Kasarda & Lindsay, 2012), the notion of 'crisis' may be a strange, if not contentious, point of departure for any discussion of the future of aviation as air transport globally is arguably in good health. Since the end of the Second World War, demand for passenger and cargo flights has increased dramatically to the point where 2.8 billion passengers and 48 million tonnes of airfreight flew around the world in 2012 (Fig. 1 and ATAG, 2012). Currently, in the region of 1,700 commercial airlines (operating over 20,000 aircraft) fly 30 million commercial flights between 3,750 airports worldwide every year (Airlines for America, 2013; ATAG, 2012). However, the spatial distribution of these services is highly uneven with the majority of flights being concentrated at major cities in the economically developed world. Crucially, however, although certain air transport markets, including those in North America and parts of Western Europe, are believed to be close to saturation, others, particularly in the rapidly developing economies of Latin America and the Middle and Far East (including most notably Brazil, the United Arab Emirates, China, Indonesia and the Philippines), are rapidly expanding,

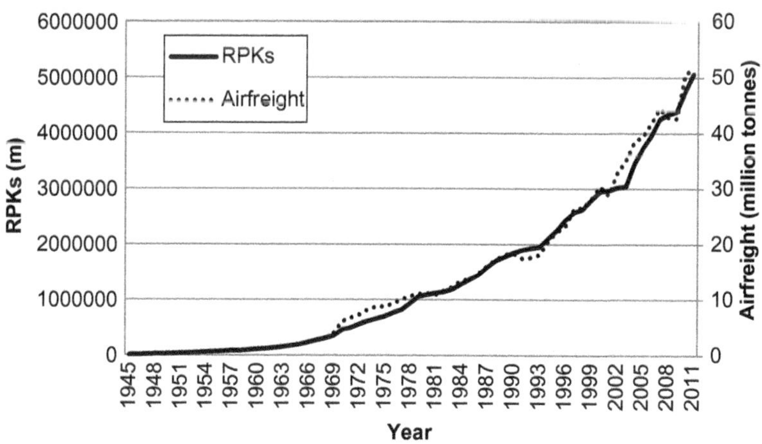

Fig. 1. Growth in Global Airline Revenue Passenger Kilometres (RPKs) and Airfreight, 1945–2011. *Source*: Data derived from Airlines for America (2013).

leading to growing demand and a net increase in flights (Boeing, 2012). Given the growth potential of these markets, the International Air Transport Association (IATA) has forecast an annual global passenger growth rate of 5.3 per cent from 2013 to 2016, by which time 3.6 billion passengers are expected to fly each year (IATA, 2012a).

The inexorable expansion of commercial air transport after the Second World War, which actually accelerated at the end of the last century thanks, in part, to global policies of airline deregulation and the emergence of low-cost carriers, appears in the eyes of industry representatives unlikely to falter in the near future. Talk of a crisis in such circles is perhaps a little premature unless it refers to a crisis of airport capacity and of routes between new emerging destinations and mature markets. In its 'most likely' scenario for the development of the European aviation market in 2035, Eurocontrol (2013, p. 15) thus foresees 1.5 times more commercial air traffic movements than in 2012. It predicts that growth will be most robust in Eastern Europe and more rapid for traffic to and from Europe than within the continent while Turkey will generate more additional flights than any other country. As John Bowen explains in Chapter 2 of this volume, the contours of the global aviation industry are being progressively redrawn as a result of changes to the regulatory environment and the global economy, but this does not constitute a crisis of aviation *per se*. A similar line of argument may be taken towards the restructuring of the industry. Following market liberalisation and the 2008 global recession in particular, a number of major carriers have collapsed and significant market adjustments have occurred in response to the competitive threat posed by low-cost carriers, the shifting balance between leisure and business travel and the withdrawal of routes and concomitant reduction in passenger numbers at certain airports. Such adjustments are, however, to be expected and do not in themselves signal any long-term crisis of demand.

Paradoxically, the apparent resilience of the global aviation sector as a whole and the growing demand for air travel in new markets has inculcated it into a further set of interconnected social, economic and environmental crises. Put simply, flying, or rather mass aeromobility, with all its systems, structures, procedures, practices and languages, has immersed itself in our everyday lives. It has shifted our conceptions of time and space, offered us new mobilities, quickened practices of globalisation, and contributed to the development of the global neoliberal order in the second half of the twentieth century (Urry, 2009). This has come to constitute over time what Urry and others suggests is a culture of 'air-mindedness', which goes beyond the departure lounge or airport terminal and which spreads its tentacles out into how we think about and structure our economy, organise our social interactions and plan our urban and rural futures (Urry, 2009, pp. 25, 36; see also Adey, 2010; Randles & Mander, 2009). As a result, how we celebrate landmark events in our lives, go on holiday, organise sporting events, choose the food we shop for at the local supermarket, keep in contact with colleagues and relatives, manage and forge business partnerships, profit from foreign markets, conduct international diplomatic relations, understand migration and communicate with one another are all increasingly shaped by our experiences, and possibilities of, air travel.

But, it is the very pervasiveness of the expansionist logic of air travel that locks aviation firmly into a set of contradictory challenges, the origins of which might not solely lie within the practices of flying, but which amount nonetheless to a transformational crisis or series of crises for aviation. Indeed, it is hard to refute that, alongside the highly contentious issues of airport expansion, location, noise pollution and quality of life for those living near airports, a further litany of charges and threats against aviation have been added to its balance sheet, from security and safety concerns, public health fears, and social injustice, to the spread of new forms of corporate imperialism, visions of ever-expanding urbanisation and 'faster living', and the threat of peak oil and our reliance on fossil fuels, which challenge the very existence of mass aviation. More significantly however, in both scientific and public discourse, aviation has been repeatedly identified as a growing contributor of carbon emissions and greenhouse gases and tied inescapably to the universal challenge of climate change.

The salient and very public link between air travel and climate change has done much since the start of the 21st century to begin to dislodge the dominant narrative of global aviation success and the economic necessity of its expansion (Griggs & Howarth, 2013a). Aviation, or specifically its capacity for further expansion as well as many of the carbon-intensive practices it supports, has been challenged as being incompatible if not wholly contradictory with government policies to tackle climate change (Anderson, Shackley, Mander, & Bows, 2005; Cairns & Newson, 2006; see Bows-Larkin & Anderson, 2013, Chap. 3; Anger-Kraavi & Köhler, 2013, Chap. 5). Commercial aviation currently accounts for around 3 per cent of all carbon dioxide (CO_2) emissions that result from human activities, but growing demand for air travel and emissions reductions in other sectors means that its total contribution is likely to increase. In the United Kingdom, the Committee on Climate Change (CCC) (2009) has predicted that commercial aviation's contribution to UK greenhouse gases will rise to 25 per cent by 2050 if aviation continues to expand. This scenario would require other sectors to reduce their emissions disproportionately if the United Kingdom is to meet its carbon reduction targets. In short, capacity constraints and congested skies might well continue to frustrate travellers and allegedly cause delays that cost the global economy billions of US dollars every year in lost productivity, but they are no longer the only game in town: climate change has challenged, or at least destabilised, the traditional boundaries of the policy and public debate.

Of course, such claims over the impact of aviation on climate change are heavily disputed. On the one hand, there are competing evidence bases and areas of uncertainty, such as the impact of contrails and aircraft emissions at high altitude on radiative forcing. On the other hand, there are ongoing debates over the ability and effectiveness of technological change to lower aviation emissions (and over what timescale this might occur), as well as the stringency of domestic and international regulatory regimes and the viability of emissions trading, be it on a regional or global scale (see Anger-Kraavi & Köhler, 2013, Chap. 5). For example, the United Kingdom's CCC in its 2009 report suggested that a 60 per cent growth in demand for air travel could be compatible with the commitment to keep CO_2 emissions from commercial aviation in 2050 no higher than they were in 2005. This apparent mismatch between higher demand and lower emissions was explained through anticipated future fleet fuel efficiency, the use of alternative fuels, and enhanced air traffic management and operational procedures (CCC, 2009); technologies, which despite industry advances, are yet to enter widespread use (see Budd & Budd, 2013, Chap. 4).

In fact, in the field of aviation, at least in mature markets, there has been a hardening of the political boundaries between rival coalitions as the 'new' politics of aviation protest has transformed campaigns against airport expansion (Griggs & Howarth, 2004). Early campaigns tended to mobilise against noise pollution and to conserve countryside and rural environments and protect the quality of life of residents living near airports or proposed sites for expansion. In many instances, protesters did not necessarily contest the expansion of aviation, but rather challenged the location of new airports or the appropriateness of the sites that had been selected for development (although the potential to broaden campaigns was always present as the campaign in the 1970s against Narita airport demonstrated (Apter & Sawa, 1984)). However, borrowing from the tactical repertoire of direct action movements and the availability of new discourses of sustainable development and the environment, local anti-airport campaigns have diversified their strategies and coupled flying to issues such as tackling climate change, advancing alternative forms of sustainable transport, challenging the limits of government decision-making, and addressing global justice, thereby forging universal campaigns against airport expansion at *any* site and indeed *against* air travel itself. In so doing, they have deepened alliances with environmental groups, anti-corporate lobbies, direct action networks and radical environmentalists, as well as trade unions, new farming movements, local authorities and celebrities, as the campaign against the third runway at London Heathrow and the construction of a new international airport on agricultural land at Notre-Dame-Des-Landes near Nantes in France have recently demonstrated.

These new coalitions against aviation expansion are, we suggest, part and parcel of the multiple crises facing the aviation industry in the second decade of the 21st century. Griggs and Howarth (2013a) argue that one of the successes of the new protests against airport expansion is the transformation of aviation from a 'tame' into a 'wicked' policy issue for government (Rittel & Webber, 1973). 'Wicked issues' are characterised by conflicting policy frames, each informed by competing evidence bases, rival definitions of problems and solutions, and antagonistic beliefs and values. They are multidimensional issues that cross traditional policy boundaries to such an extent that they are relatively immune to one-shot policy solutions and are fraught with unexpected consequences as acting on one aspect triggers negative consequences for other elements. As such, they present a complex set of interconnected economic, political and social governance challenges that require a reframing of the very issues at stake (Schön & Rein, 1995).

Against this background, aviation, or rather the issue of aviation, has become if not a 'crisis' at least an impasse for many governments and international bodies (Griggs & Howarth, 2013a). Economic liberalisation and deregulation, as well as climate change, have further fragmented and multiplied the competing logics at play in the policy sphere of aviation, surfacing the dependency of any government on an array of external stakeholders, which includes other governments, given the international regimes that regulate air travel, as well as global environmental lobbies, local resident groups, aviation environment organisations, international financial conglomerates, national airlines, low-cost carriers, aviation regulators and industry associations.

Critically, there is no broad consensus over the policy instruments, strategies and behavioural incentives that might permit governments and the aviation industry to address rising emissions. As we have suggested, this ambiguity in both policy and practice derives, in part, from conflicting interpretations of the impact aviation emissions have on both the global climate and on local noise and air quality around airports. It also stems from uncertainties surrounding the nature of aviation's contribution to the global economy; the constraints imposed by international regulatory regimes governing commercial aviation; the credibility of the predicted improvements in aeronautical technologies; the current capacity constraints of airport infrastructure; and the continuing rise in public demand for air travel (Gössling & Upham, 2009). Indeed, governments of all hues risk being caught between a current dependency on air travel and aeromobility *and* a broad recognition of the need for changes to the aviation industry and more importantly to the embedded social practices of flying (which, if not a reality for many across the globe, remain a desirable aspiration nonetheless).

It is these issues and challenges that this volume seeks to illuminate. As we suggest above, one starting point is to recognise the multiple and often contradictory crises facing the aviation industry, some of which are only partially the product of the internal contradictions of air travel. However, it is our contention that because flying has entered and shaped our social and economic practices to such an extent, it cannot remain immune to such crises or be an industry that warrants special treatment. In addition we would countenance against quick appeals to positive-sum games through which aviation expansion and carbon reductions go hand in hand, particularly given the obstacles created by the global interdependencies of the industry. Indeed, the reframing of aviation policy will necessarily be fraught with technical and political difficulties, engage multiple, and often antagonistic, coalitions, and take place across all levels of society and government. With this in mind, in the next section, we start to analyse the tensions and contestations that inform contemporary understandings of 'sustainable aviation' in industrialised western economies. In particular, we set out the contested realities, ambiguities and contradictions of pursuing (or not) sustainable futures for aviation, which contributors to this volume then go on to explore.

CONTESTED REALITIES, SUSTAINABLE FUTURES AND AVIATION

'Sustainable aviation', or rather the political battle to formulate and implement such policies, has come to increasingly dominate and structure the politics of contemporary air travel. Indeed, the very phrase 'sustainable aviation' is widely disputed – what political scientists and policy analysts often call a 'contested concept' (Walker & Cook, 2009). It has been dismissed as an oxymoron, while also being critically evaluated as an ideological move by governments and supporters of aviation expansion to ward off opposition to proposals to increase airport capacity (Griggs & Howarth, 2013a). Yet, for others, 'sustainable aviation' as a package of workable policies is already on the horizon, driven by technological improvements and by international agreements on emissions trading. The 2012 inclusion of international aviation in the European Union emissions trading scheme (ETS), for example, reduced certain elements of opposition to further expansion at London Heathrow on the grounds that increased capacity will not lead to increased emissions given the caps on emissions (Griggs & Howarth, 2013a).

These disagreements are grounded not merely in competing interpretations of scientific knowledge or rival impact assessments of policy tools. They are rooted in different webs of ethical and ideological beliefs, diverse attitudes to risk and technology, and rival narratives of the past and visions of the future (Hulme, 2009, p. xxvi). In such circumstances, the resolution of differences is not straightforward or compatible with so-called appeals to objective evidence-bases, for what we might term to be different policy frames or discourses of aviation constitute the very problems and solutions, evidence bases, and understandings of aviation under scrutiny. In other words, 'flying' is a political construct which is constantly re-constructed and brought into being by different protagonists and practices. This assertion engages us not in the exploration of the reality of air travel, but in the critical analysis of its contested multiple realities.

In this section, we explore the different realities of aviation by analysing the contentious boundaries and fault lines which organise ongoing public dialogues over the future of 'sustainable aviation'. To structure this critical assessment, we borrow from the discourse of sustainable development. Sustainable development is widely understood, in the rhetoric of the Brundtland Commission (United Nations World Commission on Environment and Development, 1987, Chap. 2, para. 1), to mean 'development that meets the needs of the present without compromising the ability of future generations to meet their own needs'. As such, it advocates that individuals, firms, public bodies and governments must not prioritise one particular need over another, but consider how actions impact both positively and negatively on economic, social and environmental outcomes across societies. Importantly, such assessments must be grounded not merely in the short-term and decision-makers must 'future-proof' policies so as not to compromise the opportunities of future generations.

Of course, any definition of sustainable development is itself highly contentious and subject to rival different interpretations, as our discussion of the meaning of 'sustainable aviation' has intimated. Equally, any assessment of the sustainability of air travel cannot evaluate its different contributions to wellbeing in isolation from one another. Yet, it must also avoid descending into narrow cost-benefit exercises, working instead, we suggest, from a broad vision of future needs and desires concerning the quality of life across societies. Here, we recognise such concerns, and we do not propose any definitive assessment of the sustainability of air travel. Rather, we deploy the widely accepted definition of the Brundtland Commission as a heuristic device to expose the competing sustainable futures of aviation and the divisions between them. Therefore, we first examine the competing interpretations of the economic, social and environmental impacts of air travel, before turning to how these different interpretations constitute multiple scenarios for the future of aviation. We now turn to the economic rationalities that inform different aviation futures.

Claims for Connectivity

The aviation industry claims to employ directly over 8 million people across the globe (ATAG, 2012). However, part and parcel of the narrative of the expansion of commercial aviation is its depiction as a 'vital' cog in the modern global economy and hence a driver of social progress. Aviation, it is often repeated, is a primary catalyst in the reproduction of the economic wellbeing of the modern nation-state as well as an agent of social progress within communities that provides more and more people with more and more opportunities for cultural interaction and exchange (Griggs & Howarth, 2013a). In its briefing on the economic benefits of aviation, the IATA typically spells out these oft-lauded strategic advantages. It first acknowledges the direct employment benefits from air transport, but then foregrounds aviation's 'essential input' into the global economy through the increased connectivity air transport networks provide. 'Greater connections' by air, it suggests, drives growth by 'providing better access to markets, enhancing links within and between businesses and providing greater access to resources and international capital markets' (IATA, 2007, p. 1). Indeed, IATA's home webpage lists the economic benefits of aviation, presenting its visitors with the claims, amongst others, that '$6.4 trillion of goods travel by air – that's 35% of all world trade by value. **Aviation delivers**' or that '3.5% of the global economy relies on aviation. **Aviation supports business**' (IATA, 2013a, emphasis in original).

However, these claims are being increasingly challenged, constituting a growing fault line in the politics of aviation. Airports have been transformed in recent years into commercial centres, with much of their profits coming not from aeronautical fees but from retail outlets and parking (Graham, 2008). Equally, it is suggested that the economic performance of aviation is falsely inflated by tax exemptions on aviation fuel and international services (a position enshrined in the 1944 Chicago Convention and the vast majority of bilateral air service agreements). In its pamphlet on aviation and climate change, GreenSkies (2005, p. 1), an international alliance of environmental organisations and citizen groups seeking to reduce the negative impacts of air travel, argues that aviation environment and citizen organisations, argues that aviation enjoys huge tax breaks and is therefore far too cheap: 'There is no tax on aviation fuel. [...] Additionally, no VAT [Value Added Tax] is paid on aviation transactions (although the majority of EU states impose VAT on domestic air travel). All this means that each year the aviation industry in the European Union receives over 45 billion [Euros] in tax concessions and other subsidies'. Of course, industry supporters dispute these claims. They point to recent increases in air passenger duties and tourism taxes which have, they conclude, resulted in air travel being highly taxed in relation to other forms of transport. The IATA portrays an industry facing 'thousands of taxes and fees on its operations and services, [...with] the revenue raised from such taxes [...] far outweighed by the economic benefits that are forgone as a result' (IATA, 2013b). But, nonetheless, Joaquín Almunia, Vice-President of the European Commission responsible for Competition Policy, announced in July 2013 a public consultation on state subsidies in aviation which would set out plans to reform operating costs at regional airports in an effort to avoid 'duplication of unprofitable airports' and the under-use of regional facilities (2013, p. 2). Indeed, the proliferation of regional point-to-point airports in Europe has resulted in a number of airports operating significantly below capacity or even being abandoned as in the case of Spain's £1.1 billion Ciudad Real Airport (which closed in 2012 three years after opening as an 'overspill' airport for Madrid Barajas) and Sheffield City and Plymouth airports in the United Kingdom. Such closures are not limited to Europe as the failure of Yangyang International Airport in South Korea aptly demonstrates (Sudworth, 2009).

Yet, campaigners against aviation do not simply argue that the environmental costs of expansion outweigh its economic benefits, but rather question the underlying benefits of increased connectivity for economic growth and wellbeing. In their 2013 assessment of aviation and connectivity, which was funded by the World Wide Fund for Nature (WWF), the Royal Society for the Protection of Birds (RSPB) and the local residents' association, Heathrow Association for the Control of Aircraft Noise (HACAN ClearSkies), the Dutch environmental consultants, CE Delft, threw doubt on the evidence-base supporting the privileged status of air travel as an 'economic catalyst', and criticised the capacity of traditional cost benefit analysis to take full account of social and environmental costs of flying. Indeed, the CE Delft report argued that 'there is no proof that extra connectivity results in economic growth' and concluded that 'studies that claim a causal relationship between expansion and growth were found to have serious methodological shortcomings' (CE Delft, 2013, p. 2).

Equally, what we earlier posited as market transitions in the global aviation industry, particularly since the 2008 economic recession, are being understood alternatively as the first signs of a transport revolution in which commercial aviation is already suffering the effects of the economic downturn, the increasing substitution of high-speed rail for air routes and the rising price of fuel (Gilbert & Perl, 2010). The cost of aviation fuel, Gilbert and Perl estimate, increased threefold from 2002 to 2006, only to double in 2008 before falling back to 2006 prices by late 2008. These rising fuel costs and an economic recession meant that the global aviation industry lost over $10 billion in 2008 (2010, p. 95). Rising prices, Gilbert and Perl go on to claim, will threaten cheap flights, lower passenger demand and contract the market for US domestic air travel by 2025 and international air travel thereafter (2010, p. 96) as flying for leisure becomes so expensive that people 'take just once-in-a-lifetime holidays involving a grand tour of another continent' (2010, p. 258).

Overall, therefore, claims for connectivity and assessments of the future economic success of aviation demonstrate the difficulties of reaching agreement over a sustainable future for aviation. Is it an industry in terminal decline or is it in transition? Should governments treat it differently from other industries as air travel triggers other forms of economic development? These very questions themselves, let alone the responses to them, cannot be divorced from other judgments such as the likelihood of peak oil and the availability of substitute forms of transport or alternative fuels, all of which inform part of the assessment of aviation as an industry facing the threat of terminal decline. We will address these issues later, but before doing so, we examine the assessment of the social impacts of aviation that structure contemporary debates.

Claims for Social Progress

As we alluded to above, flying has become intrinsically tied to our everyday lives. Air travel facilitates what certain societies now see as needs and/or aspirational goods, be it foreign holidays, round-the-year availability of fresh foodstuffs, foreign-produced consumer goods, and opportunities for new encounters, employment, and education. Cheap air travel in such instances becomes a tool to advance social mobility and to spread social welfare. Yet, at the same time, we have noted how aviation creates negative impacts on social wellbeing, such as concerns over airport security, the global transmission of infectious disease, or the continuation of forms of social injustice and inequality.

Assessing such outcomes and assigning some form of value to them rests on particular visions of society, how one delimits 'social' outcomes and the responsibility attached to aviation for the existence of such benefits or injustices. For example, within hours of aircraft being grounded in northern European airspace due to the volcanic ash cloud caused by the 2010 eruption of Eyjafjallajökull, news reports and social media sites carried stories of stranded passengers, fears of lost holidays and shortages of fresh green vegetables and cut flowers, all of which revealed our dependence on the availability of cheap air travel. Yet, these stories were quickly countered by those revelling in the delight of empty skies, the absence of aircraft noise and the return, albeit short-lived, to the social benefits of 'slow living' (see Budd, Griggs, Howarth, & Ison, 2011). Indeed, in an earlier voicing of such criticisms, Mark Ellingham, founder of the Rough Guides travel books, commented on society's addiction to 'binge-flying', claiming that 'we now live in a society where, if people have nothing to do on a Saturday night, they go to Budapest for 48 hours. We fly anywhere at the slightest opportunity, 10 times and upwards a year' (Observer, May 6, 2007).

With these rival understandings of the social outcomes of air travel in mind, we focus our attention on three contentious elements of aviation's social footprint: fairness, cohesion and social justice. The low-cost revolution in aviation, as we have stated, has opened up international air travel and exchanges to lower income groups. As Shaw and Thomas (2006, p. 209) argue, this extended opportunity for 'holidays, short breaks, visiting relatives, educational, cultural and religious exchanges' has significant consequences for 'social and spatial equality'. More importantly, they argue that the 'democratisation' of air travel has transformed 'people's desire for air travel into a consumer expectation, a norm or even a right'. Yet, these social impacts of air travel are arguably still reserved to a minority of the world's population. Even within mature aviation markets such as the United Kingdom, flying remains dominated by an affluent elite. In a 2011 survey of passengers at London airports, managerial, administrative and professional workers accounted for over 80 per cent of leisure travellers at Heathrow and almost 75 per cent of passengers at Gatwick. Even at Luton and Stansted, airports that are predominately used by low-cost carriers, managerial, administrative and professional workers still dominated, although there was evidence of a shift towards more supervisory, clerical and junior managerial or administrative workers (who accounted for 45.6 per cent of passengers at Stansted as opposed to 27.5 per cent from higher or intermediate managerial or professional employment) (CAA, 2011, p. 76).

These imbalances in rates of flying raise critical issues for any assessment of aviation's social value. Like aviation's economic impacts, its social benefits are being increasingly challenged. HACAN ClearSkies, the local resident association opposing expansion at London Heathrow airport, has argued that flying actually reproduces or exacerbates social inequalities, claiming not least that second-home owners with properties abroad are the most frequent flyers (HACAN, 2003). More broadly, the negative impact of mass tourism on local cultures and environments, which has been fuelled by the availability of flight-based package holidays, has long been recognised (Whitelegg, 2000).

Such broader impacts, well beyond the narrow confines of airports, question the claiming of air travel as a 'right'. In the first instance, the costs of aviation are disproportionately experienced by those living near airports or on flight paths through rising levels of noise pollution. Within this perspective, air travel is a practice undertaken by many at the expense of the few (although whether the 2 million people thought to be affected by aircraft noise from Heathrow airport counts as 'few' is an interesting point given that approximately 68 million passengers use the airport each year). While noise pollution may well figure as an environmental dimension of air travel, its impacts on education and community wellbeing are significant. According to a 2013 study around Heathrow, children living under flight paths take an additional two months to develop reading skills than other children (British Broadcasting Corporation [BBC], 2013). Annoyance from noise can also provoke tensions as it interrupts normal daily activities and communication (Hume & Watson, 2003, p. 57). More broadly, airport operations and expansion can threaten the destruction of village communities and impact adversely on perceptions of empowerment and social justice when local community campaigns are defeated and expansion goes ahead (see Griggs & Howarth, 2002). Of course, this latter impact may owe more to government and its consultation processes than to aviation itself.

Yet, with climate change and rising greenhouse gas emissions from aviation, the direct costs of flying are being felt by broader swathes of the globe and are impacting on other species. Campaigners have been quick to connect flying to such negative outcomes, with the local resident group, Stop Stansted Expansion, inviting the Inuit leader, Aqqaluk Lynge, to speak on its behalf in July 2007 against airport expansion and on the dangers of climate change to the Arctic ecosystem at a public inquiry into increasing capacity at the airport (Stop Stansted Expansion, press release, 21 July 2007). More recently, in a 2010 viral internet campaign, Plane Stupid, the direct action network against airport expansion, drew attention to the impact of flying on other species and their habitats. It released a short film of polar bears falling from the sky, crashing into buildings and parked cars against the background noise of planes passing overhead. The film ends with the message that 'an average European flight produces over 400 kg of greenhouse gas for every passenger ... that's the weight of an adult polar bear'.[1] But, whatever the power of such visual rhetoric in its defence of the rights of other species, it also neatly leads us to consider the environmental impacts of flying, which is our third primary fault-line within the politics of aviation futures.

Claims for Environmental Protection

As Budd and Budd in Chapter 4 (2013) show, aviation creates a range of negative environmental impacts including noise and local air pollution which can, over time, exacerbate existing health concerns and lead to a range of physical and mental health impairments (Hume & Watson, 2003). Airports also generate significant volumes of surface access traffic which contributes to local air pollution (Whitelegg, 2000, pp. 8–11). They also promote development of surrounding (often rural) areas and their presence can disrupt habitats and/or alter natural water basins.

An October 2012 report on air quality and aviation expansion in the United Kingdom, undertaken by the Laboratory for Aviation and the Environment at the Massachusetts Institute of Technology and the Energy

Efficient Cities initiative at Cambridge University, estimated that 110 people die early each year in Britain from airport emissions; a figure which was calculated to rise to 250 early deaths a year by 2030. This rise in the number of early deaths was explained by the greater use of airports, growing and ageing populations, and the increased impact of aviation emissions in what was determined to be a cleaner atmosphere in the future (Barrett, Yim, Stettler, & Eastham, 2012, p. 5). To tackle rising air pollution, the report suggested near-mitigation measures such as removing sulphur from aviation fuel, single engine taxiing, the electrification of ground support equipment and avoiding using aircraft auxiliary power units, as well as considering prevailing winds and population density in decisions to expand airports (2012, pp. 4–5).

The extent of these localised impacts, their relative weighting in decision-making, and the effectiveness of measures to offset them is one of the fault lines structuring the policy arena of commercial aviation. As Bröer discusses in Chapter 7, reducing noise pollution has been a constant demand of local residents and those living under flight paths since the introduction of jet aircraft in the 1960s. Indeed, extensive consultative machinery has been put in place to manage noise impacts on local communities. Yet, these interventions and the introduction of quieter aircraft have done little to reduce demands to lower noise pollution, which remains one of the primary causes of local resident mobilisation against airports. Direct auditory damage from aircraft is said to be rare, but with no agreed standard or decibel level at which noise impacts are deemed to become significant, it is difficult to assess noise and its impacts on levels or perceived levels of annoyance, sleep disturbance and stress (Hume & Watson, 2003). If anything, the contradictions of noise pollution have increased as scientific evidence has come to question the accepted decibel levels at which noise becomes irritating to those subjected to it.

Turning to the global environmental impacts of air travel, all stakeholders engaged in the debates surrounding the future of aviation broadly accept that aviation contributes to rising levels of carbon emissions. Fault lines exist, however, over the extent of its contribution; the relative importance of such emissions compared with those of other industries; the rate of aviation emissions growth in the short-term and medium-term; and whether or not technological developments or trading schemes can effectively reduce or offset aviation's contribution to rising carbon emissions (Bows, Anderson, & Upham, 2009; Gössling & Upham, 2009). Indeed, as we suggested above, although commercial aviation currently accounts for around 3 per cent of all carbon dioxide (CO_2) emissions that result from human activities, if current growth forecasts prove accurate, the number of annual aircraft departures could increase from 31 million in 2012 to 59 million worldwide by 2030 (International Civil Aviation Organisation [ICAO], 2013) with potentially serious global implications for the global climate.

In the United Kingdom, reducing carbon emissions from commercial aviation has been identified as one of the primary challenges facing the British government if it is to meet its commitment to reduce carbon emissions by 80 per cent by 2050 (Anderson et al., 2005; Cairns & Newson, 2006). As we have argued, the UK CCC (2009) predicted that aviation's contribution to UK greenhouses gas emissions will increase to 25 per cent by 2050 if aviation capacity continues to expand, thereby requiring other sectors to reduce their emissions disproportionately if the United Kingdom is to meet its reduction targets. Yet, two significant uncertainties hang over this estimate. First, as the Coalition government acknowledged in its 2011 scoping document, predictions of continued air travel expansion may well fail to incorporate adequately into their calculations the impacts of peak oil and rising oil costs on passenger demand given the uncertain development of alternative fuel sources.

Secondly, the radiative forcing or warming effects of aviation emissions are still not fully understood because they occur at high altitude and go beyond the impact of carbon dioxide (Lee et al., 2009). Contrails are believed to trap long-wave radiation from the ground adding to global warming, while soot and sulphate emissions are implicated in the formation of anthropogenic cirrus clouds. Yet, such impacts on global warming and cooling are far from straightforward and thought to vary by latitude, altitude and season (over the Arctic, for example, their impact may be more significant) (Jacobson, Wilkerson, Balasubramanian, Cooper, & Mohleji, 2012; Schumann, Graf, Mannstein, & Mayer, 2012; Whitelegg & Williams, 2000, p. 19). It is misguided therefore to delimit aviation's impact on global warming to its level of carbon emissions, although much uncertainty remains as to the impact of aircraft emissions at high-altitude (Lee et al., 2009; Whitelegg & Williams, 2000, pp. 19–20). Indeed, updating IPCC estimates with 2005 data as well as including aviation's impacts on cirrus cloud formation, Lee et al. (2009) concluded that aviation contributed 4.9 per cent of radiative forcing, higher than its CO_2 impacts alone.

Worldwide, growing awareness of air travel's deleterious social and environmental impacts has prompted aircraft manufacturers, airlines, and airport operators to invest in new, less carbon intensive, aeronautical technologies and adopt new operating procedures to lower aviation's carbon footprint. These include the introduction of more fuel efficient aircraft such as Airbus' A320neo (new engine option) and Boeing's 737MAX, research and development into alternative fuels, and enhanced air traffic management techniques such as continuous climb departures and precision area navigation. In Europe, the SESAR programme aims to confer significant environmental benefits through air traffic management efficiencies associated with a single European sky.

Yet, as Chapter 4 in this volume details, the effectiveness of these technologies is highly contested. While industry representatives point to a record of technological advances in producing quieter and more fuel-efficient aircraft and engines as evidence of its capacity for technological change, its detractors raise the complexities of the technological advances required to wean aviation off kerosene and the broader negative impacts that such a transformation may engender. Take, for example, the use of aviation biofuels as a substitute for conventional Jet A/A1 fuel. The European Biofuels Flightpath programme, which involves the European Commission, carriers including Lufthansa, Air France/KLM and British Airways, and biofuel producers, aims to develop a supply chain capable of producing 2 million tonnes of sustainably produced paraffinic biofuels by 2020.[2] In its 2012 report on biofuels and aviation, IATA (2012b, p. 9) speaks of a 'seamless transition to low-carbon air travel'. However, AirportWatch, a UK umbrella organisation of environmental groups and local resident groups opposed to expansion, dismisses biofuels as a 'dangerous diversion' (2011, p. 2). They question the sustainability of biofuels and draw attention to the fact that biofuel production is not carbon neutral, and the need to grow sufficient quantities of suitable feedstocks induce land grabs by speculators, lead to the destruction of forests, and compete for land against food crops. Indeed, it concludes that 'the misguided rush into biofuels will encourage us in the rich world to believe we can continue to fly – and drive – even more than we now do' (2011, p. 2).

In response to the growing environmental and public relations challenge it faces, the aviation industry has set itself a number of legally binding and voluntary targets to reduce levels of aircraft noise and greenhouse gas emissions. In 2005, a group of UK airlines, airport operators and aerospace manufacturers founded Sustainable Aviation to detail 'the collective approach of UK aviation to tackling the challenge of ensuring a sustainable future for our industry' (Sustainable Aviation, 2013). In 2009, the not-for-profit aviation association the Air Transport Action Group (ATAG), an international consortium of over 50 major aerospace companies, established a number of new aviation sustainability targets.

These included commitments to increase fuel efficiency by an average of 1.5 per cent per year between 2009 and 2020, to stabilise emissions from 2020 through carbon neutral growth and an (aspirational) target of reducing aviation emissions by 50 per cent by 2050 compared to 2005 levels (ATAG, 2013).

More recently, in June 2013, IATA agreed to the development of a global emissions trading system to cap aviation emissions from 2020. This followed the suspension of intercontinental flights from the European Union's Emissions Trading System (ETS) to allow ICAO, the United Nations regulatory body overseeing aviation, to negotiate a global agreement, given the opposition and challenges by China and the United States to the European scheme. Time will tell whether this represents a significant shift in the thinking of the aviation industry or another attempt to defend air travel from, and delay the implementation of, top-down regulation by governments and international bodies. However, if reactions to the implementation of the EU ETS are anything to go by, the determination of any global market will no doubt be highly fraught with political standoffs over the calculation of emissions across the industry, the weighting of reductions between mature and immature markets, and the lobbying of governments by powerful aerospace and aviation companies. As it stands, the EU has sought to exercise its political muscle, warning that for it to extend its exemption of intercontinental flights, ICAO must make 'adequate' progress on a global emissions deal by autumn 2013.

But, the EU ETS, and emissions trading in general, has itself been brought under scrutiny (see Chapters 3 and 5 in this volume). Critics of the scheme have warned that it is not a 'complete policy solution' and have drawn attention to its failure to take account of radiative forcing impacts and of the potential to expand absolute aviation emissions through offsetting while calling for tougher caps on aviation and rises in the percentage of permits that are auctioned rather than freely distributed to carriers (Lockley, 2011, pp. 31–32). Amidst these considerations, the price of carbon, and how far it will need to rise to bring about reductions in flying, becomes another fault line dividing the competing stakeholders in the aviation policy arena.

This cleavage over the efficacy of emissions trading exposes divisions over the limits of voluntary or self-regulation and the necessity for external or government regulation of air travel's environmental impacts (Daley & Preston, 2009). The limits of emissions trading, as well as technological change, require, it is argued, government action to lower demand for air travel, including cancelling plans for expansion, ending short-haul flights and raising taxation on airline tickets and aviation fuels (see Lockley, 2011). These measures are to go hand in hand with the broader substitution of rail for short-haul flights, greater encouragement of behavioural change and the use of technology such as video-conferencing for business meetings. Indeed, high-speed rail has been offered up to government as a ready-made 'policy solution' to the dependency on aviation, although it too has attracted criticism over its environmental credentials, as Dobruszkes and Givoni (2013) explain in Chapter 8 of this volume.

In short, this assessment of the environmental, and social and economic, impacts of aviation demonstrates the number of uncertainties, ambiguities and cleavages dividing protagonists in the aviation policy arena. At the same time, it illustrates how any assessment of sustainable futures for air travel cannot be divorced from the policy instruments at the disposal of decision-makers, different worldviews or frames, and the priorities given to different outcomes. On the one hand, any critical evaluation of the environmental impacts of air travel rests on the additional assessment of how far alternative measures can be made to lessen or regulate such impacts. On the other hand, these considerations cannot be made in isolation from one another for as the ethos of sustainable development dictates, no single priority should be privileged in any consideration of aviation's future. With this in mind, we offer a few concluding words about how these different dimensions of aviation and policy instruments are being brought together in different scenarios or aviation futures.

AVIATION FUTURES

Drawing on the work of Griggs and Howarth (2013a), we analyse four different viewpoints or scenarios for the future of aviation: post-carbon; high-modernism; market regulation; and demand management. Post-carbon scenarios predict the inevitable collapse of aviation in the near future. Echoing the 'presaging apocalypse' myth that frames may debates surrounding climate change (Hulme, 2009), continued aviation expansion is problematised as an unsustainable and ultimately detrimental practice of modern capitalism. Here peak oil, rising carbon emissions and the social stigma of flying have together allegedly triggered the first transformations in the managed decline of air transport. Typically, in an article in *The New Republic* entitled 'the future of aviation, the end is nigh' (26 April 2010), Bradford Plumer speculates on the end of mass aviation, suggesting that 'early signs of an aviation apocalypse are already upon us'. Pointing to the increasing articulation of flying as a 'social stigma', as well as peak oil, emissions trading and increasing costs of flying, he suggests the end of cheap flights and the return to the elite jet-set flying practices of the 1930s. But, in its more radical variants, advocates of post-carbon futures even portray aviation as the 'new' tobacco industry in its irresponsible corporate profit-seeking, its 'capture' of government departments (as evidenced by the absence of tax on aviation fuel) and its harmful impact on individuals, particularly on those who do not fly, but who will suffer from the severe impacts of climate change on their local environments, and finally its ultimate inability to maintain its corporate reputation within public opinion as scientific evidence mounts against it. Interestingly, in response to British Airways' launch of a new domestic flight between London Gatwick and Newquay airport in Cornwall, which coincided with the ban on smoking in public places in England in June 2007, Greenpeace placed full-page advertisements in UK broadsheet newspapers that mocked BA's '120-a-day habit' to domestic flying and suggested it was time for the airline to 'quit its dirty habit'. A large image of an aircraft trailing cigarette smoke as opposed to contrails across an otherwise clear blue sky accompanied text that stated 'only tobacco companies are as cynical, and they took decades to publicly accept the damage their products cause. With climate change we simply don't have that long'.

In contrast to this approach, high-modernist or techno-managerialist scenarios foresee continued expansion in aviation and place their faith in continual human and scientific progress (Hulme, 2009, p. 351). Proponents of high-modernism thus reject scare-mongering stories of the impending demise of aviation. On the contrary, they stress aviation's crucial role in the workings of the modern economy, presaging the dependence of advanced capitalism on the 'connectivity' offered by air transport. The challenges of aviation's impact on climate change are thus conceptualised as a set of manageable risks which given the right financial incentives can be mitigated through technical innovation and human ingenuity. Significantly, high-modernist scenarios thus marry in part with the discourse of ecological modernisation in which the promises of green technology are said to be able to ensure a 'positive sum game' in which both aviation growth and environmental protection are possible (see Hajer, 1995; Mol, 2000). In other words, aviation can continue to expand as a 'clean' or sustainable industry.

Market-regulation scenarios view air travel as central to the workings of the modern economy, but recognise that in the future flying will be subject to market-style hands-off restrictions or forms of regulation, which may see the number of flights capped or go into decline. These scenarios tend to put their faith in 'soft' regulatory mechanisms, economic instruments such as cap and trade schemes and carbon offsetting for tackling aviation's impact on climate change. Such mechanisms induce, it is claimed, forms of 'responsible agency' or indeed responsible corporate agency (Paterson & Stripple, 2010), which limit the role of government to setting overall limits to emissions and managing incentives. In fact, cap and trade models even allow for the continued expansion of aviation as the industry can continue to buy allowances from other industries or credits from 'clean energy' projects. In other words, expansion is possible if the aviation industry demonstrates its capacity to expand whilst reducing or allowing for its environmental impact. Put alternatively, it is an approach that ultimately takes account of the economic importance of aviation, enabling it to continue as a 'necessary evil', albeit as one which expands either through its internal efficiencies or less charitably at the expense of other industries.

Finally, demand management scenarios conceptualise the future of aviation as an industry under managed decline in which government takes the lead in lowering demand for air travel, encouraging new forms of sustainable transport, raising taxation on flying and imposing strict emissions criteria. Flying persists, but only as 'an option for truly urgent travel', replaced by rail and even wind-powered hybridised ships (Gilbert & Perl, 2010, p. 257). Examining measures to facilitate such transitions in the United States, Gilbert and Perl thus make much of the demand for strategic leadership by government (2010, p. 238). They point to three primary developments to launch the required transport revolution: first, the creation of a new public Transport Development Agency to 'plan, facilitate and monitor transport redesign' (2010, p. 239); secondly, the end to all existing plans and programmes for airport expansion (and road-building); and thirdly, the taxation or increased taxation of oil-based transport fuels, including aviation fuel (2010, pp. 239–247). Importantly, proponents of demand management eschew the limits of self-regulation in favour of government intervention and regulation to reduce global social and economic dependence on flying.

Of course, the boundaries between these aviation futures are porous. It would be misguided to draw too clearly defined lines or oppositions between them. But, within each scenario or viewpoint, policy instruments or social, economic and environmental impacts of aviation are given distinct meanings and attributed specific consequences. Which scenario or aviation future wins out in the end rests to a large degree on political negotiation and persuasion, and how rival coalitions are able to structure the terrain of argumentation over air travel. Interestingly, in its determination of future scenarios for European aviation in 2035, Eurocontrol (2013, pp. 10–11) determines the 'most likely' scenario to be that of 'regulated growth' of aviation, arguably more in line with our conceptions of market regulation than those scenarios of demand management, high-modernism or post-carbon worlds. These latter scenarios align respectively more with Eurocontrol's visions of a 'happy localism' with less globalisation and more trade and travel within Europe, 'global growth' through technological advance, or a 'fragmenting world' of higher fuel prices and reduced trade and transport integration. Yet, as this analysis of competing scenarios underlines, how we interpret the future of aviation rests on our broader values and beliefs concerning the challenges facing society and how governments and other stakeholders put in place and coordinate the multiple arenas in which a dialogue over the future of our society can be held. In other words, aviation futures cannot be reduced to the narrow confines of the technical effectiveness or feasibility of policy instruments; they merit a much wider dialogue, which explores how we imagine the futures of our societies and practices of government.

The challenges faced by policy makers endeavouring to facilitate a constructive public dialogue about airports and aviation are highlighted by the obstacles that have confronted successive British governments aiming to broker a workable policy settlement in this field. In 2002, New Labour's national consultation on airport expansion failed to produce the intended consensus on the future of aviation. On the contrary, it led to a hardening of the political frontiers between contending forces as the Labour government came down firmly on the side of aviation expansion. With Labour's plans also then discredited by a growing series of scientific and expert challenges, as well as the emergence of a powerful climate change coalition against aviation expansion, the Conservative-Liberal Democratic coalition reversed plans to construct a third runway at London Heathrow airport, when it came to power in 2010 (Griggs & Howarth, 2013a). But since the spring of 2012, the Coalition has faced a sustained political campaign from supporters of the aviation industry. In September 2012, it announced the setting up of the Davies Commission on airport capacity, due to report in 2015, after the next general election – a move seen by many as kicking aviation policy into the long grass. One year on, however, it appears that the Coalition may well be on the verge of reversing its opposition to airport expansion, with the Davies Commission having announced in October 2013 that its 'provisional view' is that 'we will need some net additional runway capacity in the south east of England in the coming decades' (Davies, 2013).

Like many previous governments, the Coalition has put its faith in the capacities of an expert commission to build support for a new policy settlement (Griggs & Howarth, 2013b). The Commission, led by Sir Howard Davies, the former chair of the Financial Services Authority, has from its creation repeated its commitment to 'an open and consultative approach' (Davies, 2013). In its first year, it released five discussion papers for consultation, undertook some 70 visits and meetings, held two days of public dialogues, appointed an Experts Advisory panel and established a Sustainability Reference group. Its call for short- and medium-term proposals to improve the use of existing airport capacity attracted some 75 submissions, matched by over 50 long-term options for airport expansion.[3] At the same time, it has also recognised the political reality of policy-making in the field of aviation. Howard Davies thus declared in his October 2013 speech outlining the emerging thinking of the Commission that 'our final conclusions will not please everyone, I am sure. But they will be better informed as a result of all this effort. I hope we have improved the quality of debate on what has been and remains a fraught topic'.

In fact, as Griggs and Howarth (2013b) argue, the contested evidence-bases and rival constructions of the very problems facing aviation weaken the capacity of the Davies Commission to generate a short-term political consensus in aviation policy. In reality, Howard Davies's claims that the Commission's conclusions 'will be better informed as a result of all this effort' risk falling on stony ground, leaving it open to increasing challenges over its own democratic and representative credentials. As a precursor to such challenges, Geoff Muirhead, who was the former head of Manchester Airports Group, and which were now the new owners of Stansted, resigned from the Commission in September 2013, following threats of legal action by the local resident group Stop Stansted Expansion over what the latter perceived to be his unacceptable 'conflict of interest'. (Manchester Airports Group had already submitted plans for a new runway at the airport.)

With the prospect of another round of national consultation, if the Commission enters its second phase of preparing detailed schemes for airport expansion at specific sites, it is tempting to see this first phase of the Commission as little more than a '*drôle de guerre*' or phoney war; as New Labour found out to its cost, the logic of consultation can harden antagonisms between rival stakeholders, rather than deliver conciliation and convergence on a reasonable agreement (Griggs & Howarth, 2013b). Of course, it is conceivable that a sensible compromise will be brokered, and that the main political parties, groups, movements and affected communities will eventually come to accept it. But it remains an open question as to whether governments have the appropriate tools and practices at their disposal to deliver such compromises, particularly when set against the embedded discourse of aviation expansion that has framed government policy since the Second World War (Griggs & Howarth, 2013a).

In announcing the provisional thinking of the Commission, Howard Davies suggested that the Commission may well come to settle on a high-modernist vision of aviation's future. He thus underlined the economic contribution of aviation, while continuing to argue that carbon reduction and increased connectivity 'are not irreconcilable goals'. Indeed, he announced that the work of the Commission 'so far suggests that doing nothing to address the capacity constraints in our current airport system would not be the right approach. Its likely effect would be to restrict passengers' choices and it could have unintended consequences for the efficiency and resilience of UK airports, as well as possibly leading to some flights and emissions being displaced to other countries' (Davies, 2013). Such declarations from the chair of the Airport Commission resonate with the embedded discourse of aviation as a privileged driver of economic and social progress, evoking the threats of foreign competition and the fears of over-capacity, rising passenger demand and falling choice that have framed post-war aviation expansion and the policies of 'predict and provide'. It will take new forms of political leadership and social change, accompanied by alternative visions of leisure, business and mobility, to shake off such demands (Griggs & Howarth, 2013a, 2013b).

AIMS OF THE VOLUME

This volume seeks to foster dialogue and contribute to the difficult and challenging processes of generating a new policy settlement in aviation. More specifically, it begins to unpick and critically evaluate the maze of rival demands, policy positions and institutional biases that currently structure the politics of aviation. Much has been written on the future of aviation since the foundations for the debate that were laid down 10 years ago in Paul Upham, Janet Maughan, David Raper, and Callum Thomas's (2003) seminal work *Towards Sustainable Aviation*. Yet, it is quite clear from the contributions to this volume that while progress has been made in the intervening decade, we are far from resolving the tensions that exist between economic and commercial imperatives for air transport growth on the one hand and environmental responsibilities on the other. Indeed, in the shifting global context of economic transformation, new and challenging evidence of climate change impacts, and emergent international regulatory regimes, it is timely to critically evaluate the current status and future prospects for sustainable aviation, and other aviation futures.

With these aims in mind, the volume goes on to examine the multiple fault lines of aviation politics, which have been established, albeit in a cursory manner, in this opening chapter. It asks leading names in global air transport and aviation policy research to offer their own innovative assessments of the future of commercial air travel in their particular fields. Each individual contribution thus engages with important issues of contemporary debates within aviation policy arenas. Collectively, however, they come together to refine understandings of contemporary debates across aviation policy arenas and offer a broad assessment of the prospects for change in how we reframe our understandings and practices of flight.

The volume is unashamedly multidisciplinary in its ambitions. It brings together the work of geographers, political theorists, climate scientists, economists, planning experts, sociologists and transport specialists. At the same time, it draws upon a range of comparative cases, seeking to investigate the interplay between the specific dynamics of local institutions and practices and more global or universal economic and political drivers. Importantly, many of the chapters presented here were delivered as part of a British Economic and Social Research Council Seminar Series into the politics and policies of sustainable aviation that ran from January 2011 to September 2012. This series brought together multiple policymakers, practitioners and campaigners from across the aviation policy arena. This valuable interaction with these multiple stakeholders helped polish the contributions to this volume and shape its focus – and we would like to thank all participants in this series for their valuable insights, which are too numerous to mention.

THE CONTRIBUTIONS

The book is divided into three principal sections. The first examines the scale and the scope of the contemporary sustainable aviation challenge. In Chapter 2, John Bowen examines the changing spatiality of global air service provision as the balance of aviation power shifts inexorably away from North America and Europe towards countries in the Middle East, Latin

America and China. He suggests that the 21st century aeromobile world will look very different from that of the 20th century, but concludes that the long-term growth of the industry is unlikely to be reversed. Critically assessing the impact of growing emissions from aviation given the predictions for growth in emerging markets, Alice Bows-Larkin and Kevin Anderson quantify in Chapter 3 the climate challenge facing air travel. They offer a detailed analysis of a range of different emissions and future climate scenarios, which evaluates the extent to which carbon reduction must be achieved in other industrial sectors if aviation is to be allowed to continue to expand at its present rate while avoiding catastrophic climate change and significant rises in average global temperatures. They conclude that governments and societies should ultimately question whether aviation expansion should continue, given the limited technological options for decarbonisation.

The second section of the volume, which examines different challenges or fault lines in addressing issues of sustainability in air travel, opens with further discussion of the technological options available to aviation. Chapter 4, by Lucy and Thomas Budd, documents the role of aeronautical technology in improving the environmental performance of commercial aviation. By focusing on the environmental impacts of all the stages of the air service delivery chain, from aircraft construction through routine flight operations and maintenance to eventual airframe decommissioning, they examine the potential environmental improvements and efficiency gains that may be afforded by the introduction of new technologies such as biofuels, new airframe configurations, open rotor engines and more sophisticated air navigation and flight planning procedures.

This assessment of the technological options open to aviation is followed by an examination of the mechanics of emissions trading and its applicability to aviation, the effectiveness of which is increasingly contested within aviation policy arenas. In Chapter 5, Annela Anger-Kraavi and Jonathan Köhler analyse aviation's inclusion in the European Union's ETS. By performing a detailed economic assessment of the potential impacts of ETS on the aviation sector, they conclude that contrary to many concerns, there is likely to be very little adverse economic impact on the air transport sector and suggest that, if a stable and sufficiently high carbon price can be maintained, the ETS may well provide a strong incentive for airlines to re-equip their fleets with more modern and less polluting aircraft.

In Chapter 6, Charlotte Halpern examines an area of aviation policy that is often underreported in the literatures on sustainable aviation – namely the complex interplay that exists between airport actors. Drawing on her extensive research in Europe, she shows how privatisation and the shifting interactions pattern of public and private ownership has rescaled in part the territorial dimension of airport activities which have become full-blown economic actors with high-levels of autonomy. Indeed, she posits that these transformations can be tied to the limited impact of anti-airport campaigns over the long-term development of major European hubs.

One of the long-term annoyances of airport operation has been noise pollution, which has mobilised local campaigns against aviation expansion for the last fifty years or more. In Chapter 7, Christian Bröer addresses this salient issue of noise pollution, analysing the political history of aircraft noise annoyance and its relationship to sustainability. Analysing community sensitivity to aircraft noise and noise management and mitigation measures at Amsterdam Schiphol and Zurich Kloten airports, he suggests that citizens' perceptions of aircraft noise are shaped by policies to tackle noise, leading paradoxically to an intensification of the annoyance associated with aircraft noise.

However, more recently, alongside demands to lower noise and air pollution from aviation, transport lobbyists and aviation campaigners have argued that high-speed rail as a substitute for short-haul flights offers an effective means of lowering the impact of rising aviation emissions on climate change. Frédéric Dobruszkes and Moshe Givoni in Chapter 8 examine the potential of high-speed rail in Europe and the extent to which it might confer environmental benefits over flying. They demonstrate how judgements rest for example on comparative load factors and the use which is made (or not) of runway capacity which is released by the switch to rail. They conclude that there are multiple challenges to overcome if rail integration with airports is to be more than a business opportunity for airlines, airports and train operating companies.

We close this second part of the volume with an analysis of the challenges posed by the congestion within our skies and the management both of the risks it poses and are posed to it by natural events. Peter Adey, in Chapter 9, discusses the future sustainability of the socio-technical regime of aviation. Drawing on his research on preparedness and the crisis management of European airspace that was triggered by the disruption caused by the ash cloud from the eruption of Iceland's Eyjafallajökull volcano, Adey alerts us to the inherent vulnerabilities of the air transport system to natural events. He suggests that arrangements put in place to respond to moments of crisis are not unproblematic and raise questions about state sovereignty and the airlines' suspicion of regulation or lack thereof.

The final section of the volume considers the prospects for change in the aviation policy arena and the difficulties governments face in addressing the 'wicked issue' posed by aviation. In Chapter 10, James Connelly investigates the brakes placed on policy change in aviation by the continuation of 'politics as usual' or what he terms to be the continued adherence to underlying presuppositions which guide the political and economic action of ministers and policy-makers. He proposes that these unacknowledged presuppositions constrain advances to integrate environmental concerns in aviation. Indeed, the prospects for change in aviation policies, Connolly suggests, rest on the political questioning and recognition of this set of default positions favouring aviation expansion.

The difficulties and opportunities of contesting and overcoming such 'default positions' are subsequently explored in the contribution by Ute Knippenberger. In Chapter 11, she analyses the conflicts that have accompanied the development of Germany's largest airport at Frankfurt/Main. She demonstrates how these conflicts are structured by tactical descriptions of space and conceptions of airports as elements of large technological systems. In so doing, Knippenberger reveals the necessity of situating airports and their local governance within different scales, from neighbouring cities to the large labour market stretching up to 100km from the airport.

Before turning to the individual contributions that make up this volume, we want to make one final clarification. This volume does not endorse any specific future vision or policy scenario for aviation. Its ambitions are more humble, for we hope that in some small way it will inform continued debates about the extent to which we can generate a new policy settlement in aviation. While we recognise that the benefits of air travel must not be decoupled from the negative impacts they create, each contributor offers his or her own critical assessment of the policy implications or prospects for change in their specific field or dimension of aviation futures. Ultimately it is incumbent on researchers to work with politicians, communities and industry to help inform future decision making that promotes more sustainable aviation for the collective good of individuals, companies, nations and planet Earth alike. We hope that this volume takes a step in that direction.

NOTES

1. Retrieved from www.planestupid.com/polarbears. Accessed on 22 July 2013.
2. Retrieved from http://ec.europa.eu/energy/renewables/biofuels/flight_path_en.htm. Accessed on 28 July 2013.

3. These submissions and the guidance and discussion documents written by the Davies Commission can be found at its website: www.gov.uk/government/organisations/airports-commission. Accessed 30 September 2013.

REFERENCES

Adey, P. (2010). *Aerial life: Spaces, mobilities, affects.* Chichester, UK: Wiley-Blackwell.

Airlines for America. (2013). *Annual results world airlines.* Retrieved from http://www.airlines.org/Pages/Annual-Results-World-Airlines.aspx. Accessed on June 28, 2013.

AirportWatch. (2011). *Beware aviation biofuels.* London: AirportWatch.

Almunia, J. (2013, July 3). *Introductory remarks on new state aid rules for airports and airlines, press conference.* Brussels, Speech/13/606.

Anderson, K., Shackley, S., Mander, S., & Bows, A. (2005) *Decarbonising the UK: Energy for a climate conscious future.* Manchester, UK: Tyndall Centre.

Anger-Kraavi, A., & Köhler, J. (2013). Aviation and the EU Emissions Trading System. In L. Budd, S. Griggs, & D. Howarth (Eds.), *Sustainable aviation futures.* (Vol. 4, pp. 105–126). Transport and Sustainability. Bingley, UK: Emerald Group Publishing Limited.

Apter, D. E., & Sawa, N. (1984). *Against the State: Politics and social protest in Japan.* Cambridge, MA: Harvard University Press.

ATAG. (2012). *Aviation: Benefits beyond borders.* Retrieved from www.benefitsbeyondborders.org. Accessed on June 28, 2013.

ATAG. (2013). *Facts and figures.* Retrieved from www.atag.org/facts-and-figures.html. Accessed on July 22, 2013.

Barrett, S., Yim, S., Stettler, M., & Eastham, S. (2012). *Air quality impacts of UK airport capacity.* Cambridge, MA: Laboratory for Aviation and the Environment, MIT.

BBC. (2013, March 29). *Pupils learning affected by Heathrow noise*, BBC News. Retrieved from http://www.bbc.co.uk/news/uk-england-london-21975462. Accessed on July 27, 2013.

Boeing. (2012). *Current market outlook, 2012–2031.* Seattle, WA: Boeing Commercial Airplanes.

Bows, A., Anderson, K., & Upham, P. (2009). *Aviation and climate change: Lessons for European policy.* London: Routledge.

Bows-Larkin, A., & Anderson, K. (2013). Carbon budgets for aviation or gamble with our future? In L. Budd, S. Griggs, & D. Howarth (Eds.), *Sustainable aviation futures* (Vol. 4). Transport and Sustainability. Bingley, UK: Emerald Group Publishing Limited.

Budd, L., & Budd, T. (2013). Environmental technology and the future of flight. In L. Budd, S. Griggs, & D. Howarth (Eds.), *Sustainable aviation futures* (Vol. 4). Transport and Sustainability. Bingley, UK: Emerald Group Publishing Limited.

Budd, L., Griggs, S., Howarth, D., & Ison, S. (2011). A fiasco of volcanic proportions? Eyjafjallajökull and the Closure of European Airspace. *Mobilities*, *6*(1), 31–40.

Cairns, S., & Newson, C. (2006). *Predict and decide: Aviation, climate change and UK Policy.* Oxford: Environmental Change Institute.

CE Delft. (2013). *The economics of airport expansion: Consulting on the limits of economic appraisal; illustrating the diminishing returns to connectivity. Summary.* London: WWF, RSPB and HACAN ClearSkies.

Civil Aviation Authority. (2011). *CAA passenger survey report 2011. A survey of passengers at Birmingham, East Midlands, Gatwick, Heathrow, Luton, Manchester and Stansted.* London: CAA.

Committee on Climate Change. (2009). *Meeting the UK aviation target: Options for reducing emissions to 2050.* London: Committee on Climate Change.

Daley, B., & Preston, H. (2009). Aviation and climate change: Assessment of policy options. In S. Gössling & P. Upham (Eds.), *Climate change and aviation: Issues, challenges and solutions* (pp. 347–372). London: Earthscan.

Davies, H. (2013). *Aviation capacity in the UK: Emerging thinking.* Speech delivered at the Centre for London, October 7. Retrieved from https://www.gov.uk/government/speeches/aviation-capacity-in-the-uk. Accessed on 13 October 2013.

Dobruszkes, F., & Givoni, M. (2013). Competition, integration, substitution: Myths and realities concerning the relationship between high-speed rail and air transport in Europe. In L. Budd, S. Griggs, & D. Howarth (Eds.), *Sustainable aviation futures* (Vol. 4). Transport and Sustainability. Bingley, UK: Emerald Group Publishing Limited.

Eurocontrol. (2013). *Challenges of growth 2013.* Brussels: Eurocontrol.

Gilbert, R., & Perl, A. (2010). *Transport revolutions: Moving people and freight without oil* (Rev. ed.), London: Earthscan.

Gössling, S., & Upham, P. (Eds.) (2009). *Climate change and aviation: Issues, challenges and solutions.* London: Earthscan.

Graham, A. (2008). *Managing airports: An international perspective* (3rd ed.), Oxford: Butterworth Heinemann.

GreenSkies. (2005). *Rail transport: The sustainable alternative for air travel in Europe.* London: GreenSkies.

Griggs, S., & Howarth, D. (2002). An alliance of interest and identity? Explaining the campaign against Manchester Airport's second runway. *Mobilization, 7*(1), 43–58.

Griggs, S., & Howarth, D. (2004). A transformative political campaign? The new rhetoric of protest against airport expansion in the UK. *Journal of Political Ideologies, 9*(2), 167–87.

Griggs, S., & Howarth, D. (2013a). *The politics of airport expansion in the United Kingdom: Hegemony, policy and the rhetoric of 'sustainable aviation'.* Manchester, UK: Manchester University Press.

Griggs, S., & Howarth, D. (2013b). 'Between a rock and a hard place': The coalition, the Davies commission and the wicked issue of airport expansion. *The Political Quarterly, 84*(4).

HACAN ClearSkies. (2003). *A poor deal.* London: HACAN.

Hajer, M. A. (1995). *The politics of environmental discourse: Ecological modernization and the policy process.* Oxford: Clarendon.

Hulme, M. (2009). *Why we disagree about climate change: Understanding controversy, inaction and opportunity.* Cambridge: Cambridge University Press.

Hume, K., & Watson, A. (2003). Human health impacts of aviation. In P. Upham, J. Maughan, D. Raper, & C. Thomas (Eds.), *Towards sustainable aviation* (pp. 48–76). London: Earthscan.

IATA. (2007). *Aviation economic benefits: Economics briefing.* Retrieved from www.iata.org/SiteCollectionDocuments/890700_Aviation_Economic_Benefits_Summary_Report.pdf. Accessed on July 17, 2013.

IATA. (2012a). *IATA industry forecast 2012–2016.* Geneva: IATA.

IATA. (2012b). *IATA 2012 report on alternative fuels* (7th ed.), Geneva: IATA.

IATA. (2013a). *Homepage*. Retrieved from www.iata.org/Pages/default.aspx. Accessed on July 7, 2013.

IATA. (2013b). *Taxation*. Retrieved from www.iata.org/policy/Pages/taxation.aspx. Accessed on July 17, 2013.

ICAO. (2013). *Key facts*. Retrieved from http://www.icao.int/sustainability. Accessed on June 28, 2013.

Jacobson, M. Z., Wilkerson, J. T., Balasubramanian, S., Cooper Jr., W. W., & Mohleji, N. (2012). The effects of rerouting aircraft around the Arctic Circle on Arctic and global climate. *Climatic Change*, *115*, 709–724.

Kasarda, J. D., & Lindsay, G. (2012). *Aerotropolis: The way we'll live next*. London: Penguin Group.

Lee, D. S., Forster, D. M., Newton, P. J., Wit, R. C. N., Lin, L. L., Owen, B., & Sausen, R. (2009). Aviation and global climate change in the 21st century. *Atmospheric Environment*, *43*, 3520–3537.

Lockley, P. (2011). *Aviation and climate change policy in the United Kingdom: A report for AirportWatch*. London: AirportWatch. Retrieved from www.airportwatch.org.uk. Accessed on August 4, 2011.

Mol, A. P. J. (2000). Ecological modernization: Industrial transformations and environmental reforms. In M. Redclift & G. Woodgate (Eds.), *The international handbook of environmental sociology* (pp. 138–149). Cheltenham, UK: Edward Elgar.

Paterson, M., & Stripple, J. (2010). My space: Governing individuals' carbon emissions. *Environment and Planning D: Society and Space*, *28*(2), 341–362.

Randles, S., & Mander, S. (2009). Practice(s) and Rachett(s): A sociological examination of frequent flying. In S. Gössling & P. Upham (Eds.), *Climate change and aviation: Issues, challenges and solutions* (pp. 245–272). London: Earthscan.

Rittel, H. W. J., & Webber, M. M. (1973). Dilemmas in a general theory of planning. *Policy Sciences*, *4*, 155–169.

Shaw, S., & Thomas, C. (2006). Social and cultural dimensions of air travel demand: Hypermobility in the UK?. *Journal of Sustainable Tourism*, *14*(2), 209–215.

Schön, D. A., & Rein, M. (1995). *Frame reflection: Toward the resolution of intractable policy controversies*. New York, NY: Basic Books.

Schumann, U., Graf, K., Mannstein, H., & Mayer, B. (2012). Contrails: Visible aviation induced climate impact. In U. Schumann (Ed.) *Atmospheric physics. Background-methods-trends (Research topics in aerospace)* (pp. 239–257). Berlin-Heidelberg: Springer-Verlag.

Sudworth, J. (2009, May 18). *South Korea's abandoned airports*. BBC news online. Retrieved from http://news.bbc.co.uk/1/hi/world/asia-pacific/8055857.stm. Accessed on April 20, 2013.

Sustainable Aviation. (2013). *Sustainable aviation homepage*. Retrieved from http://www.sustainableaviation.co.uk. Accessed on July 29, 2013.

Upham, P., Maughan, J., Raper, D., and Thomas, C. (Eds.) (2003). *Towards sustainable aviation*. London: Earthscan.

Urry, J. (2009). Aeromobilities and the global. In S. Cwerner, S. Kesselring, & J. Urry (Eds.), *Aeromobilities* (pp. 25–38). London: Routledge.

Walker, S., & Cook, M. (2009). The contested concept of sustainable aviation. *Sustainable Development*, *17*(6), 378–390.

Whitelegg, J. (2000). *Aviation: The social, economic and environmental impact of flying*. London: Ashden Trust. Retrieved from http://waag.co.nz/pdf/res0006.pdf Accessed on July 15, 2013.

Whitelegg, J., & Williams, N. (2000). *The plane truth: Aviation and the environment*. London: Ashden Trust and Transport 2000.

World Commission on Environment and Development. (1987). *Our common future: From one earth to one world*. New York, NY: United Nations. Retrieved from http://www.un-documents.net/wced-ocf.htm. Accessed on July 30, 2013.

CHAPTER 2

CONTINENTS SHIFTING, CLOUDS GATHERING: THE TRAJECTORY OF GLOBAL AVIATION EXPANSION

John Bowen

ABSTRACT

Purpose – *The chapter discusses two main themes: shifts in the global geography of aviation toward the developing world and several threats to the future growth of air transportation. The past two decades have witnessed a remarkable realignment of air passenger and air cargo traffic toward middle-income countries and the hubs within those countries.*

Method – *The chapter documents these shifts, drawing on analyses of airline capacity around the world in 1998 and 2012. Given more rapid population and economic growth in Asia, the Middle East, Sub-Saharan Africa, and Latin America, further such developments seem likely. However, the chapter also reviews some of the principal challenges confronting aviation in its second century. These include the higher price of oil, the political challenges involved in building new airport infrastructure (especially in rich democracies), and efforts to limit the increase in greenhouse gas emissions and other air transportation externalities.*

Sustainable Aviation Futures
Transport and Sustainability, Volume 4, 37–63

ISSN: 2044-9941/doi:10.1108/S2044-9941(2013)0000004002

Findings – *The chapter concludes that none of these challenges is very likely to reverse the long-term growth of air traffic but will instead intersect with the broader shift toward emerging markets to produce a still more complex geography of air services. The chapter further contends that the continued expansion of aviation will bring both daunting challenges to the world but also new opportunities to the low-income countries still marginalized in today's airline networks.*

Keywords: Aviation geography; economic development; climate change; oil resources; emissions trading

INTRODUCTION

On August 25, 1919, the world's first regularly scheduled international commercial flight left London's Hounslow Heath Aerodrome en route to Le Bourget Airport in Paris (Davies, 1964). The flight carried a single passenger, but it nevertheless marked the beginning of London's long reign as the world's most important international airline hub. Today, the five London area airports together have a greater combined passenger total than any other city's airport system (Table 1). And yet, the primacy of London atop the world's airline networks seems unlikely to last deep into the present century. Beijing and Dubai are two among its fast-rising rivals.

Table 1. Top Cities Ranked by Total Airport System Passenger Traffic, 2011.

Rank	City	Airports with Scheduled Traffic	Passengers (Millions)
1	London	5	131
2	New York	6	109
3	Tokyo	2	100
4	Paris	3	92
5	Atlanta	1	89
6	Chicago	3	86
7	Los Angeles	5	82
8	Beijing	2	81
9	Shanghai	2	74
10	Dallas-Ft. Worth	2	64

Sources: Airport authority websites.

Europe and the Atlantic Rim more generally are fading forces in commercial aviation and London, like many long-established air transport hubs, faces significant constraints that will curtail its future growth.

This chapter addresses these two themes: first, the shifting geography of airline transportation and second, several "clouds" that may limit the sector's growth toward mid-century – including airport congestion, a limited global oil supply, and efforts to cap aviation emissions. These two themes – geographic change and gathering constraints – intersect in interesting and important ways. On the one hand, rising oil prices and policies to make aviation more environmentally sustainable are likely to accelerate fundamental changes in the geography of air transportation; and, on the other hand, the increased constraints upon the airline industry threaten to perpetuate the marginalization of the least developed economies in the world's air transportation networks. The stakes, then, in understanding the trajectory of air transportation, both in the recent past and in the longer term future, are very high.

Perhaps most fundamentally, aviation has been singled out for its significant and growing role in global climate change. One business-as-usual forecast indicates that emissions from the aviation sector alone will exceed the entire global budget for greenhouse gases consistent with no more than a +2°C change in average temperatures by 2100 (Scott, Peeters, & Gössling, 2010). And yet any effort to constrain the sector's environmental impact must contend with the stark reality (but also perhaps the advantageous opportunity) that the growth of air transportation has hardly begun across much of the developing world.

CONTINENTAL SHIFTS IN GLOBAL AVIATION GEOGRAPHY

Much as continental drift gives rise to mountain chains and to the stature of individual peaks within them, one can think of changes in the geography of global aviation as being driven at one scale of analysis by continental shifts in economic clout. And at a different scale of analysis, those shifts give rise to towering hubs, which are perennially threatened by the erosion of their advantage. The peakedness of commercial aviation networks is, at least to some degree, a refutation of Thomas Friedman's (2005) well-worn metaphor of the "flat world." His idea of a world made flat by various technologies, especially the internet, has been widely panned in many quarters. Writing in *Foreign Policy*, one reviewer scoffed, "The world isn't flat; it just looks that way from a business class seat." (Ikenberry, 2005, p. 167). Air transportation, which interestingly is hardly mentioned at all in *The World Is Flat*, is a powerful factor giving rise to a spiky world, with London perhaps the sharpest, tallest spike of all; yet changes in the geography of air services, changes made possible by new technologies and liberalization, have unquestionably brought more even accessibility – a flatter topography – across large parts of the world.

Consider the world's busiest air passenger hubs and how their traffic volumes changed during the first decade of the 21st century (Table 2). Atlanta's Hartsfield, London's Heathrow, and Tokyo's Haneda Airport are near the top as they have been for decades, yet change during the

Table 2. Traffic Growth at Top Passenger Airports, 2000–2010.

Rank	Airport	Cargo (tonnes)		Average Annual Growth (%)
		2000	2010	
1	Hartsfield-Jackson Atlanta Int'l	79.96	89.33	1.1
2	Beijing Capital Int'l	18.95	73.95	14.6
3	Chicago O'Hare Int'l	72.03	66.77	(0.8)
4	Heathrow (London)	64.38	65.88	0.2
5	Haneda (Tokyo)	56.35	64.21	1.3
6	Los Angeles Int'l	66.37	59.07	(1.2)
7	Paris Charles de Gaulle Int'l	47.70	58.17	2.0
8	Dallas-Fort Worth Int'l	60.64	56.91	(0.6)
9	Frankfurt	49.46	53.01	0.7
10	Denver Int'l	38.33	52.21	3.1
11	Hong Kong Int'l	32.10	50.35	4.6
12	Madrid Barajas Int'l	31.77	49.84	4.6
13	Dubai Int'l	10.29	47.18	16.4
14	John F. Kennedy Int'l (New York)	32.34	46.51	3.7
15	Amsterdam Airport Schiphol	39.47	45.21	1.4
16	Soekarno-Hatta Int'l (Jakarta)	10.00	44.36	16.1
17	Suvarnabhumi (Bangkok)	29.20	42.78	3.9
18	Changi (Singapore)	28.39	42.04	4.0
19	Guangzhou Baiyun Int'l	12.80	40.98	12.3
20	Shanghai Pudong Int'l	12.10	40.58	12.9
21	George Bush Intercontinental (Houston)	34.99	40.48	1.5
22	Las Vegas McCarran Int'l	36.63	39.76	0.8
23	San Francisco Int'l	40.84	39.25	(0.4)
24	Phoenix Sky Harbor Int'l	35.86	38.55	0.7
25	Leonardo da Vinci-Fiumicino Airport	22.68	38.25	5.4

Source: Airports Council International (2011a).

decade was astonishing. Beijing surged into second place, and Dubai parlayed its advantageous intermediacy vis-à-vis the world's main transport markets to become a premiere hub for transcontinental and intercontinental travel. More surprising perhaps was the staggering growth of Jakarta where the emergence of the Indonesian middle class and the aggressive growth of several low-cost carriers combined to fuel an air travel boom. Meanwhile, the disparity in growth rates among cargo hubs was similarly striking (Table 3). Of course, in looking at these top hubs, we are seeing only the peaks.

To paint a larger panorama, this section of the chapter employs data from OAG Max, a searchable database of schedules for virtually every

Table 3. Traffic Growth at Top Cargo Airports, 2000–2010.

Rank	Airport	Cargo (tonnes)		Average Annual Growth (%)
		2000	2010	
1	Hong Kong Int'l	2,275,042	4,165,852	6.2
2	Memphis Int'l	2,486,880	3,916,811	4.7
3	Pudong Int'l (Shanghai)	798,367	3,228,081	15.0
4	Incheon Int'l (Seoul)	1,877,703	2,684,499	3.6
5	Ted Stevens Anchorage Int'l	1,805,630	2,646,695	3.9
6	Paris Charles de Gaulle Int'l	1,611,875	2,399,067	4.1
7	Frankfurt	1,704,161	2,275,000	2.9
8	Dubai Int'l	583,110	2,270,498	14.6
9	Narita Int'l (Tokyo)	1,935,263	2,167,853	1.1
10	Louisville Int'l-Standiford Field	1,516,827	2,166,656	3.6
11	Changi (Singapore)	1,704,575	1,841,004	0.8
12	Miami Int'l	1,640,609	1,835,797	1.1
13	Taipei Taoyuan Int'l	1,211,812	1,767,075	3.8
14	Los Angeles Int'l	2,040,892	1,747,629	(1.5)
15	Beijing Capital Int'l	772,415	1,551,471	7.2
16	Heathrow (London)	1,405,223	1,551,404	1.0
17	Amsterdam Airport Schiphol	1,263,584	1,538,134	2.0
18	Chicago O'Hare Int'l	1,470,988	1,376,552	(0.7)
19	John F. Kennedy Int'l (New York)	1,812,979	1,344,126	(3.0)
20	Suvarnabhumi (Bangkok)	866,638	1,310,146	4.2
21	Guangzhou Baiyun Int'l	484,293	1,144,456	9.0
22	Indianapolis Int'l	1,164,482	1,012,589	(1.4)
23	Newark Liberty Int'l	1,079,320	855,594	(2.3)
24	Haneda (Tokyo)	768,663	818,806	0.6
25	Shenzhen Bao'an Int'l	154,545	809,125	18.0

Source: Airports Council International (2011b).

airline in the world. The analyses that follow are based upon OAG data for one month each in early 1998, 2008, and 2012. These particular years are partly happenstance but they are useful because they capture the geography of airline operations in early 1998 when the price of oil was just $12 per barrel, in early 2008 not long before the near-collapse of the global economy, and now. OAG data tell us about the supply side of the industry not demand (at least not directly), but nevertheless one can learn much about aviation's trajectory by examining where airlines fly and how their operations have changed over time.

The airline industry globally grew quite rapidly between 1998 and 2008 (with total global scheduled seat capacity rising an average of 3 percent per year) but slowed somewhat during and after the Great Recession (2 percent average annual growth between 2008 and 2012). Given the severity of the economic troubles afflicting large parts of the world during this more recent period, the resilience of global aviation between 2008 and 2012 is impressive and, for those concerned with making aviation sustainable, somewhat troubling. If an economic downturn as severe as that which began in 2007 could barely stem, much less reverse, the global growth of aviation, then the prospect of democratically elected governments implementing policy measures strong enough to do so seems dubious. The Great Recession *was* evident in the sharp contraction in capacity in the US, Britain, Spain, and France among other developed countries, but the reductions in service in such markets were more than offset by gains in the rest of the world, especially middle-income countries (Table 4). In fact, the BRICs (Brazil, Russia, India, and China) and the CIVETS (Colombia, Indonesia, Vietnam, Egypt, Turkey, and South Africa) all displayed the upward (and outward) mobility that has been a defining characteristic of such categories. Consider, for example, the link between Jakarta and Surabaya, Indonesia's two largest cities. In this and other middle-income countries, per capita incomes are up and airfares have held relatively steady (especially after adjusting for inflation) because new carriers have entered the fray. Capacity on the route expanded fourfold between 1998 and 2012 as an economy class roundtrip fare fell from 9 percent to just over 1 percent of Indonesian per capita income.[1] The top carrier on the route was newcomer Lion Air, one of the many upstart low-cost carriers that have begun operations in the past two decades. The overall result is surging traffic volumes – a process that is being repeated across much of the world, with daunting implications for the environment.

Collectively, the 10 countries of the BRICs and CIVETS recorded a gain of 6.6 million seats per week between 2008 and 2012 (a 50 percent increase

Table 4. Countries Ranked by Scheduled Airline Capacity, 2012.

Rank	Country	Scheduled Airline Seats per Week (000)			Average Annual Growth (%)	
		1998	2008	2012	2008–2012	1998–2012
1	USA[a]	19,980	19,824	17,380	(3.2)	(1.0)
2	China[b]	2,052	5,780	8,844	11.2	11.0
3	Japan[a]	2,933	3,035	2,976	(0.5)	0.1
4	Brazil[b]	1,201	1,791	2,749	11.3	6.1
5	United Kingdom[a]	1,753	2,982	2,476	(4.5)	2.5
6	Germany[a]	1,647	2,505	2,294	(2.2)	2.4
7	India[b]	564	1,871	2,252	4.7	10.4
8	Spain	1,045	2,422	1,922	(5.6)	4.4
9	Australia	1,091	1,577	1,727	2.3	3.3
10	Indonesia[c]	408	1,036	1,696	13.1	10.7
11	France[a]	1,424	1,859	1,659	(2.8)	1.1
12	Italy[a]	1,063	1,758	1,595	(2.4)	2.9
13	Canada[a]	1,364	1,433	1,521	1.5	0.8
14	Russia[b]	448	940	1,267	7.7	7.7
15	Turkey[c]	321	606	1,068	15.2	9.0
26	Colombia[c]	321	336	583	14.8	4.4
30	South Africa[c]	327	472	518	2.4	3.3
32	Vietnam[c]	88	292	460	12.0	12.5
35	Egypt[c]	145	260	338	6.8	6.2

[a]G-7.
[b]BRICs.
[c]CIVETS.
Sources: Author's analysis of OAG Max (1998, 2008, 2012).

in four years). Conversely, among the traditional G-7 countries, only Canada recorded an increase in airline capacity and the group as a whole suffered a decline of 3.6 million seats per week (an 11 percent decrease). It is important to emphasize that these data relate to capacity and that the movements in actual traffic were not necessarily as great. In the United States, for instance, load factors have improved from 79.2 percent in 2006 for US carriers' system-wide operations to 82.2 percent in 2011. On the other hand, some emerging economies have been beset by excess capacity as new entrants have flooded the market. Nevertheless, the clear trend over the period under review here was one of convergence between the relative size of the airline industry in middle-income and in high-income countries. That result is unsurprising given the fairly strong relationship between the level of development in a country and the relative size of its air

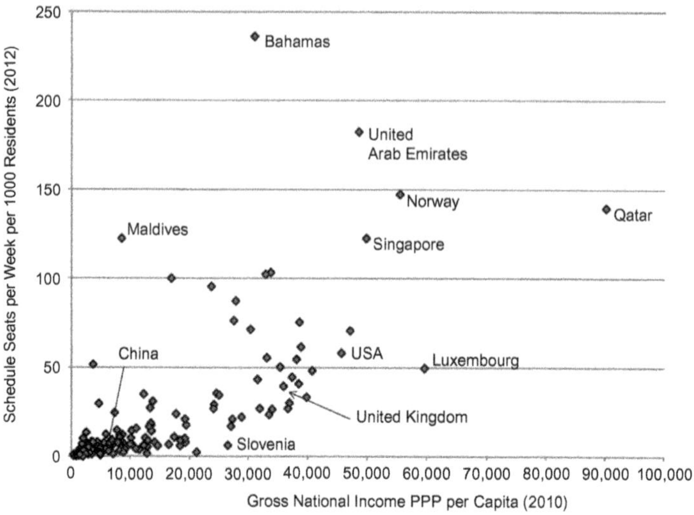

Fig. 1. The Relationship between Economic Development and Airline Capacity. *Sources*: Data from OAG Max (2012) and Population Reference Bureau (2011).

transportation sector, measured here in terms of scheduled seats per week on domestic and international routes combined (Fig. 1). The outliers on this graph are, on the one hand, those island countries and small countries with economies heavily oriented around tourism and on the other hand, countries in the shadow of dominant hubs. For instance, passenger airline capacity from Luxembourg is clearly adversely affected by the proximity of Brussels, Paris, and Frankfurt.

Fig. 1 has particular salience to the sustainability of commercial aviation because it is almost inevitable that the ratio between airline capacity (and airline passengers and cargo) and population in poorer countries is going to converge toward the high levels attained in rich countries such as Britain. As an illustration of what is possible, consider the case of the United Arab Emirates, which moved sharply along both dimensions between 1998 and 2012. The UAE is admittedly something of a special case but the very success of Dubai and Singapore (Lohmann, Albers, Koch, & Pavolvich, 2009) and other sixth freedom hubs in harnessing the power of aviation for economic development means that more countries are certainly going to try to move toward the upper-right corner of this graph.

And that is a key point that must be borne in mind in evaluating the prospect of sustainable aviation: the relationship depicted here works in both directions. Yes, people in richer countries are better able to afford the expense of air travel and air cargo and to be enmeshed in personal and business networks that encircle the globe. Yet, at the same time, good accessibility in air transportation can be a catalytic factor in economic development as has been amply demonstrated in the literature (Al Kaabi & Debbage, 2007; Brueckner, 2003; Button & Taylor, 2000; Debbage, 1999; Debbage & Delk, 2001; Graham, 2003; Hakfoort, Poot, & Rietveld, 2001; Irwin & Kasarda, 1991; Ivy, Fik, & Malecki, 1995; Kasarda & Green, 2005). In the case of Singapore, specifically, for instance, the industry accounts for more than 3 percent of gross domestic product directly and if one also counts the indirect and induced effects of the sector, its share of the Singapore economy rises to 15 percent (Oxford Economics, 2011). And of course, the true impact is broader still. There are 7,000 transnational corporations in Singapore and most would not be there if not for the linkages to the rest of the world that the airport provides. Many other countries and cities want what Singapore has.

The dynamism of the airline industry generally and the rise of middle income developing countries in particular are evident once again in a ranking of the top city-pairs for airline traffic (Table 5). On the domestic side of the ledger, the rise of Beijing–Shanghai, Delhi–Mumbai, Johannesburg–Cape Town, and the route linking Seoul and the resort island of Cheju displaced routes such as New York–Washington and Dallas–Houston from the top 15. Incredibly, there is nearly as much airline capacity between Hanoi and Ho Chi Minh City as between Chicago and New York despite the fact that the two city-pairs are nearly identical in length but the American pair has larger populations and far greater wealth. A similar point can be made with respect to Jeddah–Riyadh versus Montreal–Toronto and Guayaquil–Quito versus Lisbon–Porto despite similar populations and intercity distances for each comparison. Indeed, in each of the latter two comparisons, there was far more airline capacity in 2012 on the developing country route. These patterns reflect in part the weaknesses of alternative intercity modes in developing countries and suggest that aeromobility (Lassen, Smink, & Smidt-Jensen, 2009) is likely to become at least as much an entrenched feature of transportation in such countries as it has in most developed ones.

Among international routes, the ascendance of Dubai was evident but what is more impressive is the dominance of spans linking Asia's main hubs. The concentration of population and an increasing share of the world's wealth in this region have augmented the significance of these spans in the world's economic architecture. Not coincidentally, for instance, the

Table 5. Top Domestic and International Routes Ranked by 2012 Scheduled Airline Capacity.

Rank	Domestic City-Pair	Scheduled Seats per Week (000)	Average Annual Percent Growth 1998–2012	International City-Pair	Scheduled Seats per Week (000)	Average Annual Percent Growth 1998–2012
1	Rio de Janeiro–Sao Paulo	243	3.0	Hong Kong–Taipei	162	1.3
2	Sapporo–Tokyo	241	1.8	Jakarta–Singapore	107	3.0
3	Fukuoka–Tokyo	204	1.6	London–New York	105	(0.2)
4	Melbourne–Sydney	199	3.6	Kuala Lumpur–Singapore	100	2.3
5	Cheju–Seoul	199	6.7	Dublin–London	98	0.0
6	Osaka–Tokyo	196	4.2	Seoul–Tokyo	98	5.6
7	Beijing–Shanghai	184	7.4	Hong Kong–Singapore	83	2.8
8	Delhi–Mumbai	141	7.5	Bangkok–Singapore	79	1.6
9	Cape Town–Johannesburg	141	3.0	Geneva–London	78	6.8
10	Okayama–Tokyo	134	4.8	Amsterdam–London	77	(0.9)
11	Chicago–New York	128	(0.6)	Bangkok–Hong Kong	73	(0.5)
12	Brisbane–Sydney	107	1.1	Doha–Dubai	71	10.5
13	Barcelona–Madrid	103	0.1	Taipei–Tokyo	69	3.4
14	Hanoi–Ho Chi Minh City	101	10.2	London–Madrid	67	5.1
15	Brasilia–Sao Paulo	100	2.8	Dubai–London	66	8.9

Sources: Author's analysis of OAG Max (1998, 2012).

Asian cities featured in the table have generally moved up in rankings such as the Global Financial Centres Index (Yeandle, 2011). Meanwhile the route linking the world's two dominant financial centres – New York and London – fell slightly during the period under review. The latter result was produced by, among other factors, the force with which the Great Recession affected the US and UK economies as well as the longer term rationalization of transatlantic services. In particular, the main global alliances all enjoy a degree of antitrust immunity (ATI) in this route region, facilitating the removal of some duplicate services (Bowen, 2010, Chap. 6).

Looking beyond the individual countries and city-pairs to the broader patterns of aviation flows and their dynamism, two final points should be made regarding the continental shift that occurred between 1998 and 2012. First, inter-regional[2] traffic grew twice as fast as intra-regional traffic though more than 80 percent of seat capacity continued to be found on intra-regional routes. Growth was especially explosive on routes linking the Middle East with the rest of the world, reflecting the aggressive expansion of airlines such as Emirates, Etihad, and Qatar Airways and the concomitant enlargement of their hubs (Dubai, Abu Dhabi, and Doha, respectively). Capacity expansion was also robust, albeit from a small base, on inter-regional routes to Africa, Latin America, and Asia as population and economic growth spread further into the periphery, propelling new patterns of air service.

Second, the average stage length of scheduled flights continued its long-term climb. The flown distance of the average flight (weighted by the number of seats) increased from 1,482 kilometers in 1998 to 1,717 kilometers in 2012. The greater length of flights is testament to the increased range of commercial aircraft so that journeys that might once have required multiple stops en route can now be completed nonstop. In 1998, there were just five regularly scheduled routes globally with a nonstop distance of at least 12,000 kilometers. By 2012, there were 30 (Fig. 2). These passenger flights and similar ultra-long-haul freighter routes, such as the nightly FedEx nonstop from Shanghai to Memphis, are examples of what the geographer Eric Sheppard has called "wormholes" in the global economy (Sheppard, 2002). They warp space, drawing distant places, near and facilitate the elaboration of increasingly dense, extensive global production networks. For instance, FedEx Boeing 777-200 Long Range Freighter services permit factories to remain open several additional hours per day in Asia and still make the cutoff for same-day flights to the United States (Bowen, 2012). Further, economic globalization, increased immigration, and the retreat of

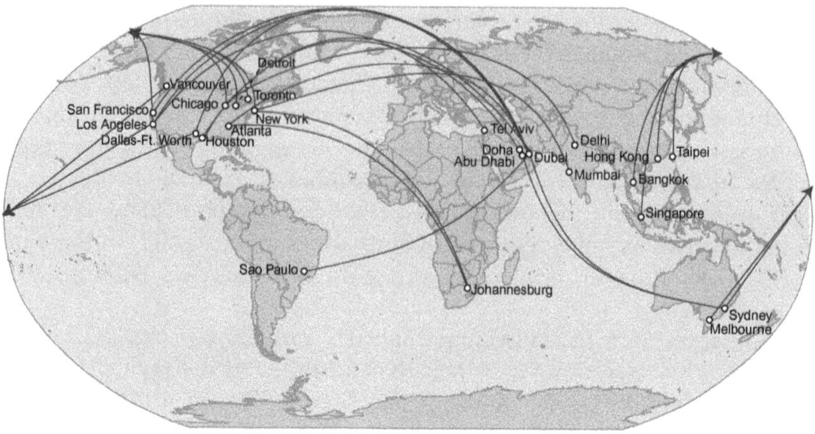

Fig. 2. Ultra Long-Haul Passenger Routes, 2012. *Note*: The routes each have a nonstop distance of at least 12,000 kilometers. There are no such routes to and from Europe partly because of that region's relative proximity to other populated landmasses. *Source*: OAG Max (2012).

the pleasure periphery have all conspired to increase demand for longer journeys (Bowen, 2010, Chaps. 8 and 9). Whatever the reason, the modest lengthening of airline services is still another factor militating against environmental sustainability in the world's transport systems since longer stage lengths limit the degree to which air transportation can be substituted for by more environmentally friendly modes such as rail. Beyond journeys of more than three hours by train, even high-speed rail becomes highly uncompetitive with aviation (Jorritsma, 2009).

THREE CLOUDS ON THE HORIZON

Despite its long record of nearly continuous growth, the aviation world seems to be entering a new era of restraint. Or to be more precise: some regions of the world are entering a new era of restraint, while others seem bound to surge upward. In this section, three limits on the future of air transportation are evaluated from a spatial perspective: global oil supplies, the provision of airport capacity, and efforts to limit the sector's role in climate change. As suggested near the beginning of the chapter, these challenges intersect in intriguing ways with broader geographical shifts in aviation flows.

Global Oil Supplies

While oil price hikes have undoubtedly had devastating effects upon the airline industry's profits and contributed to the demise of individual carriers, the overall volume of air traffic does not seem especially sensitive to the price of oil (Fig. 3). Airline traffic, whether measured in passenger-kilometers or tonne-kilometers has risen in times when oil prices rose rapidly and when they fell. Of course, the resilience of the industry might be overstated. Commercial aircraft cost up to $300 million US dollars apiece and so airlines have a strong incentive to keep them in the air even when oil prices are high. That inertia limits the ability of carriers to pass on short-term oil price hikes to travelers and shippers, which in turn limits the degree to which traffic tumbles when oil prices soar.

The experience of the price hike in 2008 *was* somewhat different in that there was a discernible dip in traffic following the all-time high oil spike of

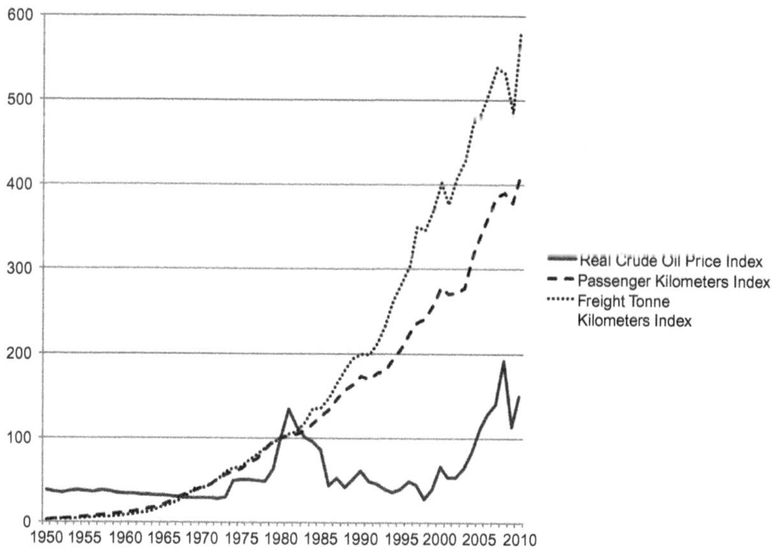

Fig. 3. Airline Traffic and Global Oil Prices. 1980 = 100. *Sources*: Energy Information Administration (2012b), ICAO (2012) and various issues of ICAO Journal.

$147 a barrel in July 2008. The surge in oil prices and concurrent economic downturn in the global economy in late 2008 claimed many carriers, including Aloha Airlines, ATA Airlines, and Skybus in the United States, Zoom in Canada, and Oasis in Hong Kong (Arnott, 2008). Around the time, the notion of peak globalization was advanced as a socioeconomic corollary of peak oil (Curtis, 2009). Evidence of that broader phenomenon included the new popularity of in-sourcing (i.e., bringing formerly outsourced manufacturing activities back to a firm's home base) as a response to high international transportation costs (Sirkin, Zinser, & Hohner, 2011).

What of the future? There has been much talk and controversy about "peak oil." For some oil producing regions, such as the North Sea, peak oil output has already passed. As for whether peak oil production globally has been passed or is drawing near, there is much less certainty (US GAO, 2007). There is, however, broad consensus that supplies of *cheaply* produced oil probably have peaked. As a result the world has become progressively more dependent on resources that are more technically difficult or situated in regions with higher production and transportation costs or greater political risks. Accordingly, US Energy Information Agency's baseline scenario is premised on a nominal oil price of $230 per barrel (or $135 in real 2010 dollars) in 2035 (Energy Information Administration, 2012a). As the supply of cheap oil begins to wane and prices rise, it is expected that across many sectors in the global economy, there will be a transition toward alternative energy sources. Because it is inherently more difficult to find viable alternatives to easily portable, energy-dense fossil fuels, however, the transportation sector generally will comprise a larger share of total consumption. And in turn, it is likely that aviation will claim an outsized share, since an airplane is even less suited to alternative fuels such as electricity than, for example, an automobile. Nevertheless, the growth of air transportation over the nearly 40 years since the 1973 Arab Oil Embargo makes it seem unlikely that global oil supplies will present an insurmountable hurdle to continued aviation expansion through the 21st century.

Instead, the increase in global oil prices and potentially dramatic changes in where oil comes from will have enormous effects on the geography of civil aviation. What is coming is a massive shift in wealth from oil importers to oil exporters and with that shift will come changes in the geography of air travel and air cargo. Consider the case of Angola, which has rapidly emerged as a top source of African oil. In 2008, there were just 45 flights and 9,000 seats per week from the capital Luanda. By 2012 there were 140 nonstops and 26,000 seats per week including new nonstops links to Beijing, Dubai, Madrid, São Paolo, among other cities. The rise of the Middle East super-connectors (*The Economist*, 2010) like Emirates and Etihad has also been fueled by the abundant petrodollars in that region. More such developments seem likely.

Airport Infrastructure

The oil-fueled expansion of aviation infrastructure is readily evident in the Middle East. Between 1998 and 2012, approximately 60 airport infrastructure projects (new terminals, new runways, or wholly new airports) with a cost of at least $500 million USD were completed across the world, and a disproportionate share was located in the Middle East, including three of the ten most expensive (Fig. 4). Yet it manifestly was not just the Persian Gulf which witnessed the massive airport build-out during the period under review. The year 2012, for instance, witnessed the construction of new airports in, among other places, Quito, Doha, and Kunming. The new airport infrastructure is testament not just to the growth of aviation but also to the hopes with which the sector is invested, especially in developing economies, as noted above.

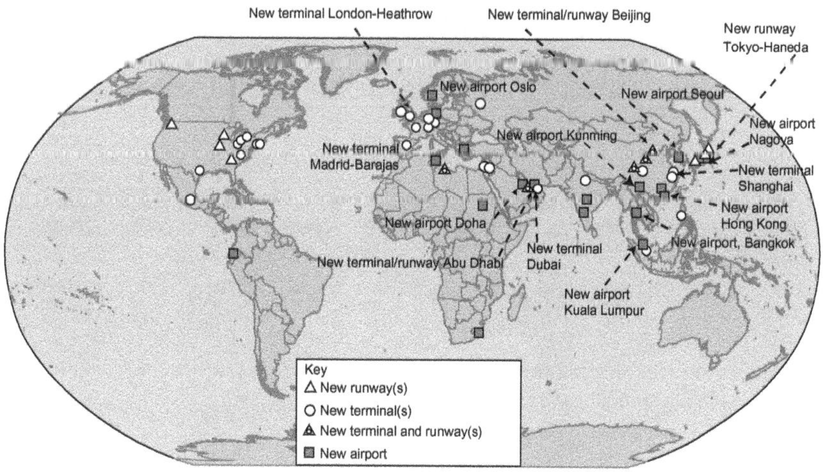

Fig. 4. 60 Most Expensive Airport Infrastructure Projects, 1998–2012. *Note*: The 15 most expensive are labeled. *Sources*: Bowen and Cidell (2011) and trade journal reports.

Nevertheless, there are reasons for skepticism that airport capacity will keep pace with demand, particularly in Western democracies. As the Greater London area illustrates, airport expansion is controversial and democratically elected governments have displayed a growing reluctance to simply build as much capacity as the airlines demand. The decisions of the coalition government in Britain not to permit the construction of additional runways at Heathrow, Stansted, and Gatwick (see Fig. 5) comprise a stark break with the practice across much of the world of building-to-forecast (*The Economist*, 2012). And now the opposition to the expansion of those airports has enveloped a new target: London mayor Boris Johnson's proposal for a new airport in the Thames Estuary (BBC, 2012). And yet the pressure in the opposite direction is fierce, too. Business groups in particular have lamented the erosion of global accessibility and the resulting loss of jobs in Southeastern England. While it is true that London is connected nonstop to more destinations than any other city in the world, Paris and Frankfurt both have more nonstops to Latin America, Africa, and the Middle East, and Frankfurt has a wide advantage in Asian destinations (Table 6).

Meanwhile, even when democracies do commit to the construction of new infrastructure, the complexities involved often produce very long lags

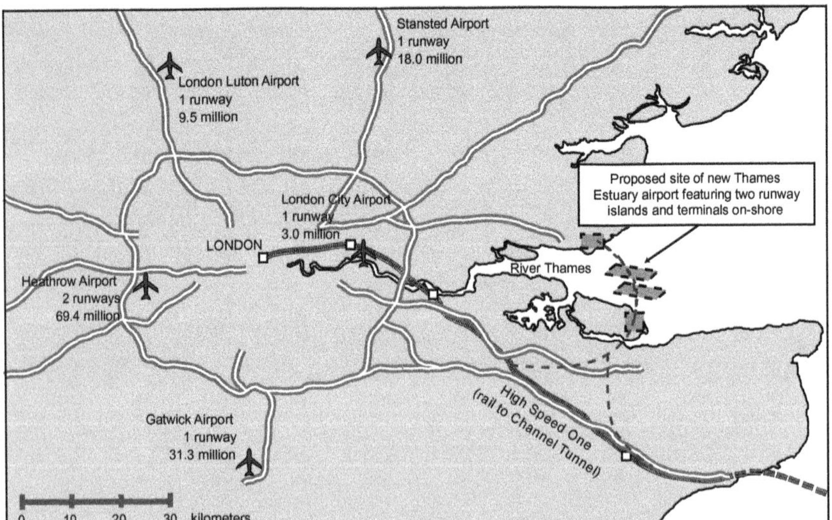

Fig. 5. London Area Airports. *Note*: The figures shown are for traffic flow in 2011. *Sources*: Bowen (2010) and airport authority reports.

Table 6. Cities Ranked by Global Nonstop Scheduled Air Passenger Links, February 2012.

City	Number of Nonstop Destinations								
	Domestic	International							Total
		Africa	Asia	Europe	Latin America	Middle East	North America	Southwest Pacific	
London	13	33	25	172	17	13	30		303
Paris	36	59	22	115	20	14	21		287
Frankfurt	13	30	33	112	21	18	23		250
Moscow	83	7	46	90	5	11	6		248
New York City	94	6	10	39	48	8	10		215
Atlanta	145	3	2	15	45	1	2		213
Amsterdam		21	20	117	18	12	20		208
Chicago	146		6	16	20	2	11		201
Beijing	110	5	36	21		5	9	3	189
Istanbul	31	16	26	83	1	27	5		189
Dubai	1	26	58	48	2	40	8	4	187
Dallas-Ft. Worth	138		2	4	33	1	4	1	183
Houston	115	1	1	5	53	2	5		182
Munich	15	14	13	101	5	9	11		168
Rome	24	14	13	91	7	11	7		167

Source: Author's analysis of OAG Max (2012).

between authorization and completion. It is striking how much more rapidly Beijing's Terminal 3 was completed compared to Heathrow's T5 (Bowen & Cidell, 2011): nearly nineteen years passed between the selection of a design for T5 and its opening, versus just over four years between the same milestones in Beijing. That is not to endorse the Chinese style of government, but the comparative ease with which less than fully democratic emerging economies can build new infrastructure is one more factor contributing to the continental shifts discussed above. As with the global oil supply, then, airport infrastructure is not likely to be a systemic constraint on the future growth of aviation but rather a factor reshaping the geography of air transport.

Alternatives to Stem Aviation Externalities

If airport infrastructure and the global oil supply are unlikely to seriously constrain the global growth of aviation, then are air traffic volumes to spiral ever upward? Malaysian low-cost carrier AirAsia's tagline "Now everyone can fly" reminds us of the near universal desire for movement. And yet the notion of a world in which everyone flies is worrying – even terrifying – from the perspective of climate change. Aviation is one of the fastest growing sources of greenhouse gas (GHG) emissions; the International Civil Aviation Organization (ICAO) has forecast that by 2050 aviation emissions will grow to 3.5 to 4.5 times their 2006 level and will comprise 20 percent of all GHG emissions compatible with holding average temperature increases to less than 2°C (ICAO, 2010); and, as noted at the start of this chapter, forecasts to the end of the century point to a much larger aviation share. All of which raises the question of whether the desire for mobility can be satisfied without watching aviation emissions and other externalities climb inexorably?

Different Modes

There are of course several ideas for permitting greater mobility while stemming those externalities (Sgouridis, Bonnefoy, & Hansman, 2011). Here, four are briefly assessed: alternate modes, alternate aviation technologies, alternate fuels, and finally emissions trading.

Among the various modal alternatives, rail has attracted most attention. Western Europe, Japan, and increasingly China have extensive high-speed rail (HSR) systems that can accommodate many short-haul and medium-haul journeys, and new links are being added – including now a newly proposed route between London and Birmingham to be completed by 2033. Unsurprisingly, more than half of all HSR construction underway in 2012 was in middle-income countries, though almost all of the new lines were being built in China (UIC, 2012). The proliferation of HSR in such markets is important as a means of preventing the technological lock-in and political inertia that favors the continued dominance of aviation in most developed countries, especially the United States.

Yet for developing countries as a whole, the extent of existing and planned HSR systems is small compared to the aviation network. Of the 17 megacities (metro areas with more than 10 million people) in developing countries, only three – Beijing, Shanghai, and Istanbul – are already served by HSR and current construction projects will not change that number. Of course, all of the mega-cities are already served by numerous airlines.

Further, high-speed rail has at most a limited capacity to divert anything but short-haul traffic from the skies. Indeed, high-speed rail may serve only to facilitate the movement of traffic to key international gateways for ever longer transcontinental and intercontinental journeys. In South Korea, for instance, the Korea Train Express or KTX opened its first segment in 2004 and since then has been extended to most of the country's main urban areas (Fig. 6). On some key domestic routes, especially Seoul to Busan (which once ranked among the busiest airline routes in the world), air traffic has tumbled (Suh, Lee, Yang, & Ahn, 2005), but overall, air traffic on routes to, from, and within Korea has climbed significantly. The KTX system is intended to eventually extend to Incheon International Airport near Seoul which will only heighten the complementary rather than competitive relationship between air transportation and high-speed rail in that country.

Of course, there are alternate modes beyond rail. Given the challenges that lie ahead, particularly in light of the geographic shifts described above, perhaps something radically different may be required. In the early 19th century, the railroad and the steamship made their commercial debut and then in the early 20th century came the airplane and the automobile. Perhaps the early 21st century has its own novelties in store.

Different Aviation Technologies

In the meantime, if airplanes are to remain intrinsic to global mobility, there is ample scope for the making them more efficient. Boeing, for

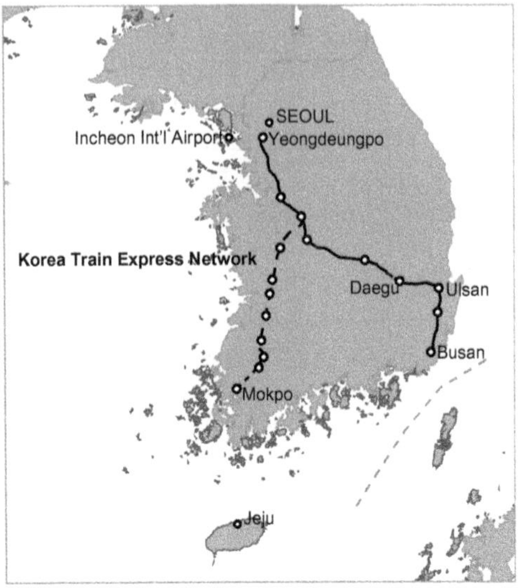

Korea Air Services (Scheduled Seats per Week)

	1998	2008	2012
Seoul-Busan	65.5	40.1	38.7
Seoul-Daegu	16.8	1.6	2.2
Seoul-Mokpo	5.0	0.0	0.0
Seoul-Jeju	39.9	85.2	101.9
All South Korea domestic routes	531.9	444.7	467.0
All South Korea international routes	182.5	504.1	550.5
All South Korea routes	714.4	948.8	1,017.5

Fig. 6. South Korea Air and High-Speed Rail Services. *Sources*: Wikipedia (2012) and OAG Max (2012).

instance, recently launched the newest version of the 737. The 737 MAX is intended to deliver 18 percent lower per passenger costs (and correspondingly lower fuel consumption, emissions, and noise) than comparably sized aircraft in service today (Boeing, 2012a). Interestingly, the largest single order for the 737 MAX has come from Indonesian low-cost carrier Lion Air (Boeing, 2012b), mentioned earlier as the top player on the Jakarta-Surabaya route. And the largest single order for the Airbus A320 Neo – which promises a similar efficiency gain – has come from AirAsia (Airbus, 2012) – showing the power of new carriers in emerging markets to shape future aviation technology.

The broader point, though, is that whatever efficiency gains the world's air-framers, aircraft engine makers, and even air traffic control authorities are able to attain will likely be passed on to consumers in the form of lower fares and air freight rates, which in turn will induce more traffic to take to the skies (Sgouridis et al., 2011). Further, commercial aircraft are designed to last about 40 years, greatly slowing fleet renewal. Similar points can be made regarding air traffic control and ground operations. Consequently, a technologically improved air transportation system will not necessarily show much improvement from a sustainability perspective. For instance, in their simulation of future aviation emissions, Sgouridis et al. (2011) found that even if the fuel efficiency of new aircraft was to improve at 2.5 percent per year between 2008 and 2024 – a rate well above the historic trend – total aviation CO_2 emissions would be only 4 percent lower in 2024 than in a baseline scenario assuming 1 percent per year aviation efficiency improvements.

An alternative route toward sustainability is to change the fuel used by commercial aircraft. There has been tremendous interest in biofuels recently, yet biofuels are laden with nearly as many externalities as conventional petroleum-derived fuels (Rye, Blakey, & Wilson, 2009). Ideally, a biofuel would not adversely affect the world's water supply, not require the use of arable land, not contribute to deforestation, incur minimal use of fossil fuels itself, grow rapidly while absorbing about as much carbon dioxide as would be emitted when the fuel derived from it is burned, and do so at a cost competitive with fossil fuels. So-called third-generation biofuels offer considerable long-term hope but the technical and commercial obstacles to attaining the required scale are formidable. One estimate is that jatropha, a third generation feedstock, if planted only in areas too marginal for conventional crops, would require 3.2 percent of the world's land area in order to yield enough fuel to supply half of global aviation requirements – and would only produce a 0.78 percent reduction in global net CO_2 emissions from all sources (Rye et al., 2009). On the other hand, biofuels obviously have advantages beyond environmental sustainability including reduced dependence on foreign, politically volatile oil sources.

With respect both to new airplanes and new fuels, efforts to improve the sustainability of commercial aviation must contend with a pronounced technology lock-in that surrounds the status quo (Kivits, Charles, & Ryan, 2010). Manufacturers, airlines, and airports have all made very expensive long term decisions based upon certain aircraft technology parameters (e.g., Jet A fuel burning engines, wingspans of up to 80 meters). Creating a consensus around alternative (especially radically different) parameters will be difficult, especially when the governments of developing countries in which aviation growth is now concentrated seem weakly committed politically to compelling such changes.

The Role of Government: Emissions Trading

In fact, government intervention to promote aviation sustainability is most evident in developed countries, especially Europe. From 2012, most airline operations to, from, and within the European Union have been subject to the EU's pioneering emissions trading system (or ETS). Specifically airlines have had to acquire sufficient allowances to cover their CO_2 emissions. In 2012, allowances equivalent to 97 percent of airline emissions during the 2004–2006 period were issued, most for free. A small share of the allowances were auctioned and it was estimated that when the bill comes due for 2012 allowances, airlines will owe EU governments a total of about €505 million (Butterworth-Hayes, 2012). From 2013, the cap on aviation emissions will be lowered slightly to 95 percent of historic emissions and the share allocated for free will be cut and so the financial cost to airlines will grow.

The great advantage of such a scheme is that it relies upon the market to encourage greater efficiency. Indeed, airlines that greatly improve their efficiency can sell their allowances to other carriers, potentially profiting handsomely during this period when most allowances are handed out for free. Further, there might be important synergies between a system like the EU ETS and more rapid progress on biofuels.

Yet the application of the EU ETS to aviation, especially international aviation, has been fraught with controversy. In particular, international airlines and foreign governments have reacted furiously to the new system, attacking the application of the ETS to the entire length of interregional flights to and from the EU, rather than just the portion within EU airspace. And governments have charged the system violates international conventions on the taxation of jet fuel used in aviation. Russia, specifically, threatened to restrict trans-Siberian overflights or to levy higher taxes on such routes – routes which play a critical role in making the world smaller (Kramer, 2012). For its part, the European Commission has defended its actions partly by pointing to the *inaction* of the International Civil Aviation Organization, the UN body responsible for addressing aviation emissions under the 1997 Kyoto Protocol.

In February 2012, 23 countries meeting in Moscow signed a declaration opposing the inclusion of international airlines in the EU ETS (Kramer, 2012). It is worth noting that all four of the BRICs signed the declaration as did the United States and Japan. This clash between the EU and emerging markets will be an interesting test of the continental shift discussed above.

TOWARD THE COMET'S CENTENNIAL

Fifty years ago, commercial jet services began with the entry into service of the de Havilland Comet. It is ironic perhaps that that first Comet flight traveled south from London to Johannesburg and by 1953 the early network of jet services stretched across both Africa and Asia, emerging markets that have figured prominently in the foregoing discussion of continental shifts in contemporary aviation (the first Comets lacked the range to attempt transatlantic flight). When the Comet first flew, the United Kingdom was among the world's largest economies. No more. Looking toward mid-century, emerging economies in particular will have markedly increased their share of world population and the global economy. What might those geographic trends mean for the airline industry?

Table 7 presents a simple, and it is fair to say, simplistic forecast of air traffic in the 2050, using the Population Reference Bureau's 2050 population projections by country, each country's current ratio of airline capacity to population, and some simple rules for how that ratio might change in the future. Almost certainly, the rankings and totals will prove to be different, maybe very different, but there is little doubt that in the coming

Table 7. A Projection of Top-Ranked Countries by Airline Capacity, 2050.

Rank	Country	2012 Scheduled Seats per Week (000)	Country	2050 Scheduled Seats per Week (000)
1	United States	17,380	China	40,561
2	China	8,844	United States	27,726
3	Japan	2,976	India	12,769
4	Brazil	2,719	Brazil	6,111
5	United Kingdom	2,476	Indonesia	5,619
6	Germany	2,294	Philippines	4,319
7	India	2,252	United Kingdom	4,252
8	Spain	1,922	Turkey	4,075
9	Australia	1,727	Spain	3,953
10	Indonesia	1,696	Malaysia	3,934
11	France	1,659	Russia	3,678
12	Italy	1,595	Germany	3,585
13	Canada	1,521	Australia	3,375
14	Russia	1,267	Mexico	3,192
15	Turkey	1,068	Japan	3,136

Source: The estimates for 2050 are based on 2012 capacity volumes, population estimates for 2012 and 2050 from the Population Reference Bureau (2011), and GDP per capita estimates for 2010 and 2050 from HSBC (2012). The ratio of per capita airline capacity to per capita GDP for each country was projected into the future with the ratio for today's richest countries falling slightly (based on expectations of higher fuel costs, major airport capacity constraints, etc.) and the ratios for poorer countries converging towards those of wealthier countries with the degree of convergence dependent on each country's forecast rate of economic growth.

decades, the very decades when climate change and the role of aviation within that conundrum are likely to become an ever more pressing issues, the locus of the global airline industry will shift farther away from Europe and other developed economies.

That is not entirely bad. Across much of the developing world, airline networks and the benefits they convey are sparse. For instance, historically, there have been few air services east-to-west across Africa. Instead, international services in Africa have tended to be north-south, often linking former colonies to past European powers. While such routes are still important to the region's airline networks, there has been a striking recent proliferation of routes between West Africa and East Africa, especially from Nairobi and Addis Ababa. These new routes illustrate the flattening of aviation accessibility referred to earlier. And yet, the paucity of links – the lack, for instance, of any nonstop connection between Lagos and Dar es Salaam – points to how much further this region's airline networks must develop to even approach the connectivity available in Europe. Or consider South Sudan, the world's newest country. As of March 2012, it had only the most skeletal of an air services network. As that network fills out, the country will gain new access to opportunities in tourism, export-oriented industrialization, the circulation of migrants and their intellectual and financial capital, and so on. Preventing the (further) development of aeromobility in such contexts would limit the growth of aviation externalities, but also limit the growth of these economies – unless viable mobility alternatives are identified and nurtured.

And almost certainly, as South Sudan, the rest of Sub-Saharan Africa, and the broader developing world are woven into an increasingly dense global network, new routes will be forged to older, somewhat eroded peaks such as London. The continued importance of those peaks will give those who control them considerable power to shape the trajectory of aviation and its externalities. Yet by the Comet's centennial that power will have been diminished and the capability and responsibility to effect a change in course will have shifted toward the kinds of places the Comet passed over.

NOTES

1. Based on annual per capita gross national income figures available from the World Bank (2012), current best fare information available on airline websites, and Jakarta Post (1998).

2. Regions defined as Africa, Asia, Europe, Latin America, Middle East, North America, and Southwest Pacific.

REFERENCES

Airbus. (2012). AirAsia orders 200 A320neo aircraft. Press release dated 23 June. Retrieved from http://www.airbus.com/presscentre/pressreleases/press-release-detail/detail/airasia-orders-200-a320neo-aircraft/. Accessed on October 12, 2012.

Airports Council International. (2011a). Passenger traffic 2010 final. Retrieved from http://www.airports.org. Accessed on February 4, 2012.

Airports Council International. (2011b). Cargo traffic 2010 final. Retrieved from http://www.airports.org. Accessed on February 4, 2012.

Al Kaabi, K., & Debbage, K. (2007). Air passenger demand and skilled labor markets by US metropolitan area. *Journal of Air Transport Management*, *13*(3), 121–130.

Arnott, S. (2008, August 30). Zoom and bust. *The Independent*, p. 48.

BBC. (2012, January 18). Thames Estuary airport plans to be examined. Retrieved from http://www.bbc.co.uk/news/uk-politics-16606212. Accessed on October 12, 2012.

Boeing. (2012a, May 2). Boeing designs advanced technology winglet for 737 MAX. Press release. Retrieved from http://boeing.mediaroom.com/index.php?s = 43&item = 2246. Accessed on October 12, 2012.

Boeing. (2012b, February 14). Boeing, Lion Air finalize historic order for up to 380 737s. Press release. Retrieved from http://boeing.mediaroom.com/index.php?s = 43&item = 2129. Accessed on October 12, 2012.

Bowen, J. (2010). *The economic geography of air transportation: Space, time, and the freedom of the sky*. New York, NY: Routledge.

Bowen, J. (2012). A spatial analysis of FedEx and UPS: Hubs, spokes, and network structure. *Journal of Transport Geography*, *20*(6), 419–431.

Bowen, J., & Cidell, J. (2011). Mega-airports: The political, economic, and environmental implications of the world's expanding air transportation gateways. In S. Brunn & A. Wood (Eds.), *Engineering earth: The impacts of megaengineering projects* (pp. 867–887). New York, NY: Springer.

Brueckner, J. K. (2003). Air traffic and urban economic development. *Urban Studies*, *40*(8), 1455–1469.

Butterworth-Hayes, P. (2012, April 4). ICAO: The best hope for settling carbon warfare. *Aerospace America*.

Button, K., & Taylor, S. (2000). International air transportation and economic development. *Journal of Air Transport Management*, *6*, 209–22.

Curtis, F. (2009). Peak globalization: Climate change, oil depletion and global trade. *Ecological Economics*, *69*(2), 427–434.

Davies, R. E. G. (1964). *A history of the world's airlines*. London: Oxford University Press.

Debbage, K. G. (1999). Air transportation and urban-economic restructuring: Competitive advantage in the US Carolinas. *Journal of Air Transport Management*, *5*, 211–221.

Debbage, K. G., & Delk, D. (2001). The geography of air passenger volume and local employment patterns by US metropolitan core area: 1973–1996. *Journal of Air Transport Management*, *7*, 159–167.

Energy Information Administration. (2012a). *Annual energy outlook 2012*. Retrieved from http://www.eia.gov. Accessed on August 30, 2012.

Energy Information Administration. (2012b). Domestic crude first purchase price by area. Retrieved from http://www.eia.gov. Accessed on March 2, 2012.

Friedman, T. (2005). *The world is flat: A brief history of the twenty-first century*. New York, NY: Farrar, Straus and Giroux.

Graham, A. (2003). *Managing airports: An international perspective*. Burlington, MA: Butterworth-Heinemann.

Hakfoort, J., Poot, T., & Rietveld, P. (2001). The regional economic impact of an airport: The case of Amsterdam Schiphol Airport. *Regional Studies*, *35*(7), 595–604.

HSBC. (2012). The world in 2050: From the top 30 to the top 100. HSBC Global Research. Retrieved from http://www.research.hsbc.com. Accessed on August 2, 2012.

ICAO. (2010). *ICAO environment report 2010: Aviation and climate change*. Quebec, Canada. Retrieved from http://www.icao.int/environmental-protection/Pages/EnvReport10.aspx. Accessed on October 30, 2012.

ICAO. (2012, January 6). Strong traffic growth in 2011 reflects improved global economic climate. Press release PIO 28/11.

Ikenberry, G. J. (2005). Review of Thomas Friedman's the earth is flat: A brief history of the twenty-first century. *Foreign Affairs*, *84*(5), 167–170.

Irwin, M. D., & Kasarda, J. D. (1991). Air passenger linkages and employment growth in US metropolitan areas. *American Sociological Review*, *56*(4), 524–547.

Ivy, R. L., Fik, T. J., & Malecki, E. J. (1995). Changes in air service connectivity and employment. *Environment and Planning A*, *27*(2), 165–179.

Jakarta Post. (1998, January 23). Garuda hikes fares, p. 1.

Jorritsma, P. (2009). Substitution opportunities of high speed train for air transport. *Aerlines Magazine*, *16*(43), 1–4.

Kasarda, J. D., & Green, J. D. (2005). Air cargo as an economic development engine: A note on opportunities and constraints. *Journal of Air Transport Management*, *11*(6), 459–462.

Kivits, R., Charles, M. B., & Ryan, N. (2010). A post-carbon aviation future: Airports and the transition to a cleaner aviation sector. *Futures*, *42*, 199–211.

Kramer, A. E. (2012). 23 countries join to fight carbon rules. *International Herald Tribune*, *24* (February), 18.

Lassen, C., Smink, C. K., & Smidt-Jensen, S. (2009). Experience spaces, (aero)mobilities and environmental impacts. *European Planning Studies*, *17*(6), 887–903.

Lohmann, G., Albers, S., Koch, B., & Pavolvich, K. (2009). From hub to tourist destination: An explorative study of Singapore and Dubai's aviation-based transformation. *Journal of Air Transport Management*, *15*, 205–11.

OAG. (1998). *OAG Max* [CD-ROM]. OAG and Downers Grove, IL.

OAG. (2012). *OAG Max* [CD-ROM]. OAG and Downers Grove, IL.

Oxford Economics. (2011). *Economic benefits from air transportation in Singapore*. Retrieved from http://www.benefitsofaviation.aero/Documents/Benefits-of-Aviation-Singapore-2011.pdf. Accessed on October 12, 2012.

Population Reference Bureau. (2011). *World population datasheet 2011*. Retrieved from http://www.prb.org. Accessed on March 2, 2012.

Rye, L., Blakey, S., & Wilson, C. W. (2009). Sustainability of supply or the planet: A review of potential drop-in alternative aviation fuels. *Energy & Environmental Science*, *3*, 17–27.

Scott, D., Peeters, P., & Gössling, S. (2010). Can tourism deliver its "aspirational" greenhouse gas emission targets. *Journal of Sustainable Tourism*, *18*(3), 393–408.

Sgouridis, S., Bonnefoy, P., & Hansman, R. J. (2011). Air transportation in a carbon constrained world: Long-term dynamics of policies and strategies for mitigating the carbon footprint of commercial aviation. *Transportation Research A*, *47*, 1077–1091.

Sheppard, E. (2002). The spaces and times of globalization: Place, scale, network, and positionality. *Economic Geography*, *78*(3), 307–30.

Sirkin, H. L., Zinser, M., & Hohner, D. (2011). Made in America, again: Why manufacturing will return to the US. bcg.perspectives. Retrieved from https://www.bcgperspectives.com. Accessed on October 12, 2012.

Suh, S. D., Lee, J.-H., Yang, K.-Y., & Ahn, B.-M. (2005). Effects of Korea Train Express (KTX) operation on the national transport system. *Proceedings of the Eastern Asia Society for Transportation Studies*, *5*, 175–189.

The Economist. (2010, June 5). Rulers of the New Silk Road. *The Economist*, pp. 75–77.

The Economist. (2012, February 4). Hub caps. *The Economist*, p. 61.

UIC [International Union of Railways]. (2012). *High speed lines in the world.* Retrieved from http://www.uic.org/IMG/pdf/20120701_a1_high_speed_lines_in_the_world.pdf. Accessed on October 29, 2012.

US GAO [Government Accountability Office]. (2007). *Crude oil: Uncertainty about future oil supply makes it important to develop a strategy for addressing a peak and decline in oil production.* GAO-07-283. Retrieved from http://www.gao.gov/new.items/d07283.pdf. Accessed on March 2, 2012.

Wikipedia. (2012). *KTX network map in 2011.* Retrieved from http://www.wikipedia.org/Wiki/Korea_Train_Express. Accessed on March 2, 2012.

World Bank. (2012). *World databank.* Retrieved from databank/worldbank.org. Accessed on October 12, 2012.

Yeandle, M. (2011, March). *The global financial centres index 9.* Retrieved from http://www.zyen.com/GFCI/GFCI%209.pdf. Accessed on August 15, 2012.

CHAPTER 3

CARBON BUDGETS FOR AVIATION OR GAMBLE WITH OUR FUTURE?

Alice Bows-Larkin and Kevin Anderson

ABSTRACT

Purpose/approach – *Despite the high profile of climate change rhetoric and the carbon intensive nature of flying, policies for controlling* CO_2 *from aviation remain at odds with global commitments on climate change. Taking a carbon budgeting approach to compare future aviation scenarios with the scale of necessary emission reductions demonstrates the extent of this contradiction. The significant potential for ongoing aviation growth contrasts with the need to curb substantially global* CO_2 *emissions across* all *sectors. For even a 50:50 chance of staying within the 2°C threshold, emission pathways imply around a 75% cut in absolute emissions by 2050 (from 1990 levels). Set against this, aviation's* CO_2 *emissions are expected to grow by between 170% and 480% over the same period, and they could feasibly be higher still.*

Originality/findings – *For the international community to be serious about its climate change commitments, moral and ethical concerns need to be considered alongside technical and economic issues. It is timely to*

Sustainable Aviation Futures
Transport and Sustainability, Volume 4, 65–84

ISSN: 2044-9941/doi:10.1108/S2044-9941(2013)0000004003

question whether expansion of an industry with few technological options for decarbonisation is a reasonable way to gamble with our future.

Keywords: Climate change; aviation; carbon budgets

AVIATION POLICY WITHIN THE CLIMATE CHANGE DEBATE

Since the early 2000s, debates within academia and wider civil society have increasingly drawn attention to the impact of aviation on the climate. Initially, most of the industry, government and the NGOs paid greater attention to noise and local air pollution, with climate change not really emerging as an aspect of environmental concern until the early 2000s. Prompted in part by the UK's Aviation White Paper of 2003 (DfT, 2003) and discussions within the EU on tackling rising greenhouse gas emissions from the aviation sector, a wider set of academics and stakeholders began to seriously analyse the potential conflict between ongoing aviation growth and reducing greenhouse gas emissions.

Historically, international transport (aviation and shipping) was not covered under the Kyoto Protocol's obligations for individual nations (or the EU as a block) to cut greenhouse gas emissions in the period 2008–2012. Instead, the International Civil Aviation Organisation (ICAO) and International Maritime Organisation (IMO) were charged with dealing with these emissions. Specifically, the Kyoto Protocol stated:

> The Parties included in Annex I shall pursue limitation or reduction of emissions of greenhouse gases not controlled by the Montreal Protocol from aviation and marine bunker fuels, working through the International Civil Aviation Organisation (ICAO) and the International Maritime Organisation (IMO) respectively. (UN, 1997)

This 'special' treatment arose primarily because responsibility for the emissions released within international airspace (and waters) is arguably not as easy to attribute to a particular nation as it is for emissions from the energy sector, for instance. For the United Kingdom, which was at that time in the process of developing its own Climate Change Act, this relegated emissions from international aviation to a low priority when devising policy measures to mitigate CO_2.

Despite the United Nations Framework Convention on Climate Change (UNFCCC) putting the onus on ICAO to drive mitigation within the international aviation sector, only marginal progress has been made. It was an absence of progress allied with broader academic inquiry into carbon budgets and emissions from the energy system (including transportation), that highlighted the importance of mitigating emissions from *all* sectors, including aviation (Anderson et al., 2008; Bows & Anderson, 2007). This debate resonated with broader concerns on climate change, with the EU subsequently proposing the inclusion of aviation within its emissions trading scheme – significantly to overcome ICAO's failure to adequately address aviation's rising emissions. Moreover, as discussion regarding the form of the policy continued, the EU surprised many by including within its final proposals not only flights within and between EU nations, but all flights departing from or arriving in EU member states.

The success or otherwise of including aviation in the EU's emissions trading system (EU ETS) depends heavily on the carbon price of tradable permits; a price subject to considerable and ongoing fluctuation. As the debate raged regarding how it may work in practice, many concluded that including aviation within the EU's trading scheme, whilst a welcome policy step, it would likely have little impact on CO_2 emissions (Bows & Anderson, 2008; Scheelhaase & Grimme, 2007). One analysis suggested that it would take a carbon price in excess of €300 per tonne before a meaningful signal would emerge through ticket price rises, if passed directly onto the customer (Bows & Anderson, 2007). Yet in 2013, the price remained below €10 per tonne, a level that would have an insignificant impact on aviation CO_2. In any case, even before the policy had a chance to get off the ground, the EU chose to suspend plans to include non-EU nations until late in 2013. Ostensibly this delay is to provide an opportunity for ICAO to discuss the implementation of its own global scheme at its 2013 General Assembly. Whilst the EU describe the suspension as a '*gesture of goodwill in support of an international solution*'[1] it has coincided with a series of legal challenges from nations outside the EU, who expect costs to rise for their aviation industries as a result of this legislation.

So where does all this leave CO_2 emissions? The short answer is 'rising' and largely unmitigated. Any changes to the level of CO_2 emissions from the aviation sector are unlikely to be impacted significantly by existing climate change policies. Aviation industries within nations outside of the EU, particularly those in industrialising countries, are growing rapidly, with obvious implications for both arrivals and departures within the EU's member states. Closer to home, debate is rife around airport expansion to support a UK economic resurgence. To quantitatively explore what this current situation may imply for climate change, the following sections assess the CO_2 emissions from aviation, how they have been changing and how they relate to the broader energy-system and climate change context.

CLIMATE CHANGE TARGETS

The term 'dangerous climate change' is widely associated with avoiding the global mean surface temperature rising by more than 2°C above pre-industrial levels. Whilst the term 'dangerous' is clearly subjective, analysis within the academic literature highlights the risks associated with temperatures going beyond this threshold and, in addition, indicates that even below 2°C there will be damage to existing ecosystems and changes to weather patterns with implications for food and water security. Every year policy and decision makers, along with science advisors and other experts come together at the 'Conference of the Parties' (COP) to negotiate how to minimise the prospects of experiencing 'dangerous climate change', with the ultimate objective of agreeing both a global cap on CO_2 and other greenhouse gas emissions along with the measures to do so. To this end, many governments have endorsed the Copenhagen Accord to '*hold the increase in global temperature below 2°C, and take action to meet this objective consistent with science and on the basis of equity*'; an Accord subsequently reaffirmed at later 'COPs' and specifically by the G8 nations in the Camp David declaration of May 2012.

It is in translating a global 2°C temperature target into global, regional and national emission-reduction pathways that nations and sectors can gain an understanding of the rate and scale by which their emissions need to reduce. However, such analysis is inevitably clouded by a range of uncertainties that stem from both what is meant by '2°C' in addition to interpretations of how readily this can be achieved through mitigation across different nations, regions and sectors. Ultimately, the detail of the debate stems from the choice of 'cumulative emissions budget', how rapidly it is assumed global emissions will reach a peak, the rate of possible mitigation and how global emissions should be apportioned to individual nations, regions and sector (Anderson & Bows, 2008).

The range in cumulative emissions arises from two principal issues: the choice of climate change model or models; and the different probabilities associated with avoiding 2°C. Importantly, the higher the probability of remaining below 2°C, because greenhouse gases are long-lived, the more constrained (smaller) will be the 'carbon budget' (i.e. the cumulative

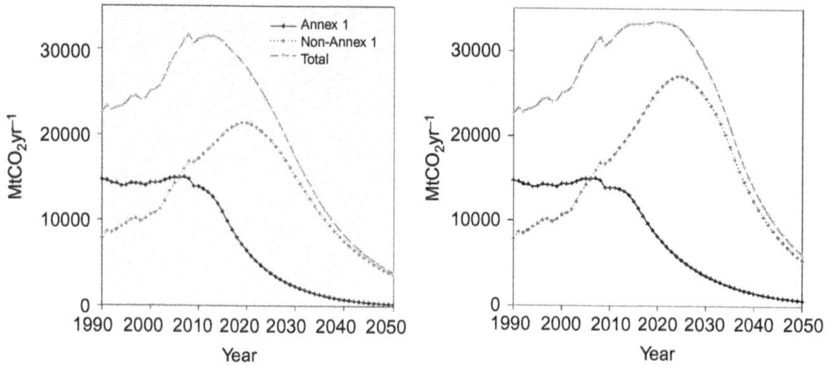

Fig. 1. Two Emission Pathways Commensurate with a 37% (Left) and a 50% (Right) Chance of Breaching the 2°C Threshold, Based on Research Published by Anderson and Bows (2011).

quantity of greenhouse gases will be lower) (Meinshausen et al., 2009). Subsequently, for a given carbon budget, the earlier emissions start to reduce (reach a peak) the more steadily they can decrease in the longer term and still remain within the chosen budget. In contrast, given the failure to yet initiate any reduction in the rate of growth in emissions, the situation is such that even if emissions are left to rise for just a few years, the post-peak reduction rate will have to increase significantly. Framing the issue within a cumulative emissions context (carbon budgets) highlights the need for short-term cuts to emissions. Moreover, in relation to aviation, it illustrates why there is such concern regarding the industry's plans and prospects for growth over the coming decade or more. As an example, Fig. 1, based on analysis published by Anderson and Bows (2011), presents two 'whole economy' cumulative emissions budgets and global pathways. The figure also divides global emissions into Annex 1 and non-Annex 1 nations, commensurate with a 37% chance and a 50% chance of staying below the 2°C threshold.

AVIATION TRENDS

Against a backdrop of the need to radically and urgently curb global emissions if the international community is not to renege on its 2°C commitments, emissions from the aviation sector continue to rise rapidly. This sector is not alone in this regard. Indeed many sectors within industrialising nations have emissions growing very rapidly. What has been interesting about the aviation sector in particular is its continued high rate of growth, notwithstanding the recent recession, within industrialised nations. For instance, prior to the economic downturn, the CO_2 emissions from international aviation were growing at a rate of around 6% per year in the United Kingdom – far in excess of the rate of national GDP growth (Bows, Anderson, & Upham, 2008). There are many reasons for why this is the case, not least of which is the deregulation of the industry which allowed low-cost carriers to gain a significant share of the market, arguably upholding growth rates that may have otherwise become somewhat depressed (Bows, Anderson, & Mander, 2009). In addition and despite the economic downturn, new markets continue to emerge between richer nations and their industrialising counterparts, as many countries, such as China, continue to have buoyant economies. Understanding the drivers behind aviation growth, in addition to how markets have developed and changed around the world, provides some insight into how emissions from this sector may change in the future. Furthermore, although it is true to say that many industrialised nations have enjoyed high rates of growth until the very recent past, the story in the United States is somewhat different. Given their share of flights globally, this situation also deserves greater attention.

To understand the importance of different nations' contributions to aviation emissions, it is useful to compare their respective levels of emissions associated with both domestic and international flying. Figs. 2a and 2b illustrate how the CO_2 produced by domestic aviation was divided amongst nations in 1990 and 2010 respectively, while Figs. 3a and 3b repeats the analysis but for international flights.

Comparing Figs. 2 and 3 highlights the very different distribution of nations' emissions depending on whether it is the domestic or international flights that are under scrutiny. Firstly, the United States dominates the domestic flying market, with a larger share of CO_2 than all other nations put together. What is all the more striking is that this dominance remains in 2010, even after the United States has witnessed a real reduction in its domestic aviation emissions of 20% compared with 1990. Nevertheless, China, as a new entrant, is rapidly gaining share with an estimated annual growth rate of 15% per annum between 1990 and 2010 according to the IEA statistics. Such an unprecedented growth is understandable when considered in relation to China's extensive land mass (similar in area to the United States), its very high rate of economic growth and its rapidly industrialising population of over 1 billion citizens. However and despite such phenomenal growth rates, domestic flying in the United States continues to

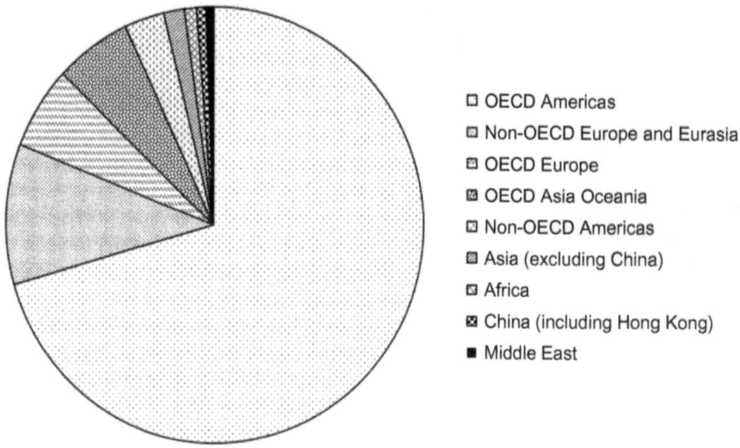

Fig. 2a. CO_2 from Domestic Aviation According to International Energy Agency Statistics, Split by Region, 1990.

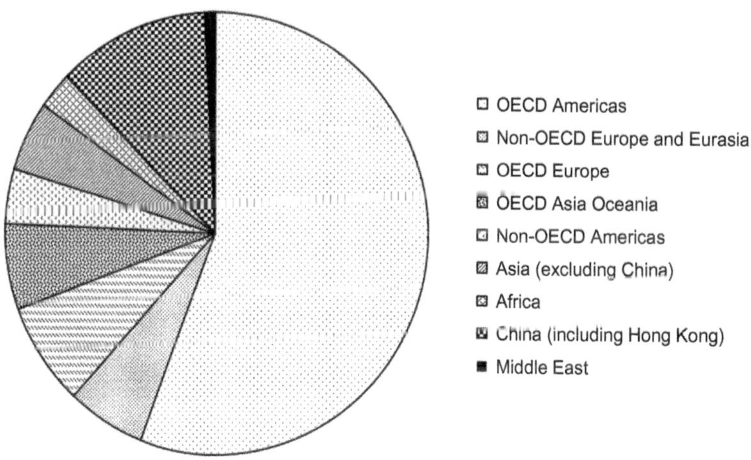

Fig. 2b. CO_2 from Domestic Aviation According to International Energy Agency Statistics, Split by Region, 2010 (total 286 $MtCO_2$).

dominate the rate of change in CO_2 from global domestic aviation; with the US decline reducing the rate of global domestic aviation growth to just 0.1% per annum from 1990 to 2010. Looking to the future, if the United States share continues to decline, attention is likely to focus on China and

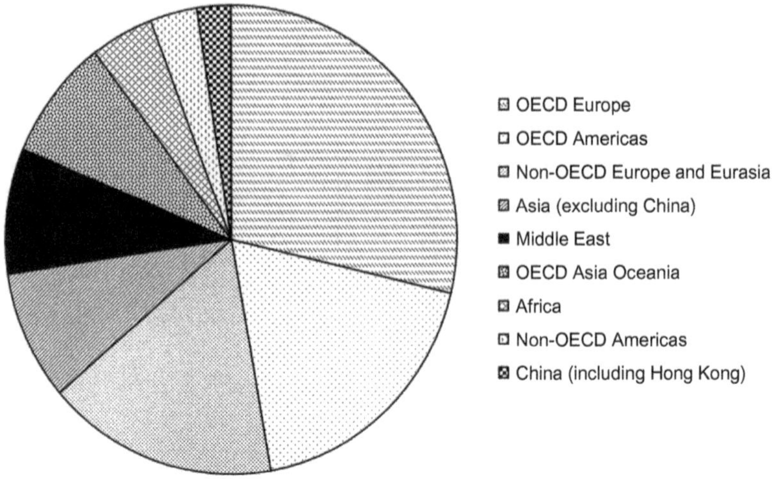

Fig. 3a. CO_2 from International Aviation According to International Energy Agency Statistics, Split by Region, 1990.

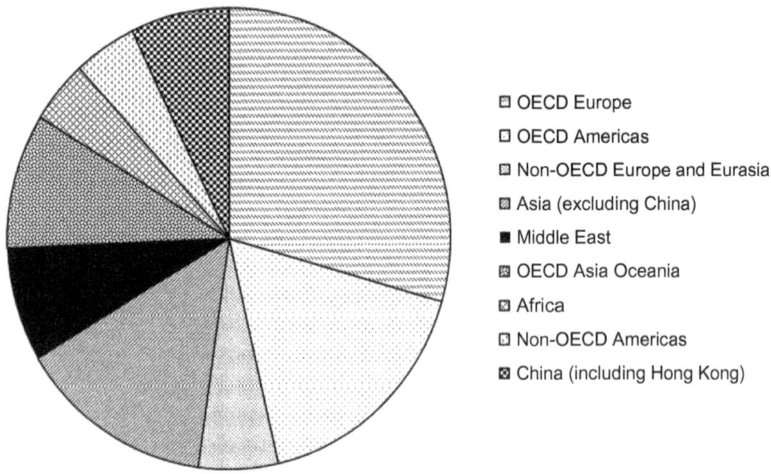

Fig. 3b. CO_2 from International Aviation According to International Energy Agency Statistics, Split by Region, 2010 (total 455 $MtCO_2$).

other large (in terms of land area) emerging economies for their greater influence on domestic aviation CO_2. A potential and perhaps probable outcome of this shifting landscape is that CO_2 emissions from domestic aviation will start to grow more rapidly again. Given the huge potential for expansion of domestic aviation within the emerging economies of China and India as well as in many nations within Africa, future growth could well exceed historical precedents.

In contrast to domestic aviation within the United States, CO_2 emissions from international flights to and from the United States have been rising at around 2.5% since 1990. So whilst the United States aviation industry is often considered to be mature, its CO_2 emissions from international flights have nevertheless increased by 64% in 20 years. A similar picture is observed in OECD Europe, where international flying has contributed to the largest emission increase in absolute terms, with an extra 60 $MtCO_2$ emitted in 2010 than in 1990.

The uncomfortable conclusion from this review of aviation trends is that there is a significant opportunity for high and probably rising levels of growth in both international flying across all nations and domestic flying within industrialising nations. Indeed, if Chinese citizens with their 0.2 flights per person per year shift towards the average of their US counterparts (2.2 flights), then there would be more flights by Chinese citizens than there are currently for the entire global population (>2 billion passengers movements).

Whilst there is a very large potential for growth in terms of flights and passengers over future decades, as discussed above, it is reasonable to question whether or not high growth rates in terms of flights and passengers necessarily translate into high growth rates in terms of CO_2 emissions. Theoretically at least, with the right combination of energy efficiency and alternative fuels, this does not have to be the case, and certainly it is such changes that other sectors intend to employ to decarbonise and cut their emissions in absolute terms. Technological, as well as operational opportunities for mitigation that can more than offset the rate of passenger and flight growth are not, unfortunately, readily available and prevalent within the aviation industry. Nevertheless, it is worth considering what the available measures are and understanding why they cannot deliver the rates of emissions reductions commensurate with the 2°C target.

MITIGATION OPPORTUNITIES FOR THE AVIATION SECTOR

Engine Technology

Minimising fuel consumption has always been a concern for the aviation industry, given its high proportion of the cost of operating a flight.

Consequently, technological developments have been significant over the previous decades, leading to very efficient high bypass high pressure-ratio gas turbine engines dominating the existing fleet. However, opportunities for continuing to significantly improve these engines diminish as time goes on. Whilst reducing fuel burn from engines will likely continue incrementally, large scale improvements to the conventional technology are unlikely, requiring fundamentally different aero-engine designs if engine emissions are to radically reduce. The open-rotor engine certainly offers some hope for moderate level improvements, but has limited application as it is much noisier than a turbo-fan, requiring major development if it is to meet noise standards. The other difficulty that is applicable to any development that cannot be retrofitted to an existing aircraft, is the time it takes for new technology to penetrate the global fleet. It will take many years before any new and radically more fuel-efficient designs delivered noticeable fleet-wide improvements. Given the rate at with which emissions can be cut in absolute terms is critical to avoiding a 2°C temperature rise, this point is extremely important.

Airframe Design

The materials used to construct an aircraft and its design and shape have a significant impact on fuel burn and hence CO_2 emissions. Both offer opportunities for some improvement in terms of fuel efficiency, but again the rate of change cannot deliver sufficiently rapid reductions to meet 2°C climate objectives. For instance, the newest aircraft increasingly incorporate composite materials, with the latest aircraft designs from both Boeing and Airbus replacing around 50% of the aluminium within the structure with composite materials. These materials are around 20% lighter than the conventional metal alloys, supporting the drive towards improved fuel efficiency. Whilst this technological development plays an important role, it is only as these newer aircraft replace the existing fleet, assuming the older aircraft are retired, that significant benefits will be gained.

The same argument applies to aircraft shape and design, although there are also infrastructure issues to consider in this case. Whilst various unconventional airframes exist, they are either in early prototype form or have uses only in niche markets, in the military, for instance. The blended wing-body aircraft is an example of this. One reason for remaining wedded to conventional aircraft designs is that the global airport infrastructure is specifically set-up for them. Even the Airbus A380, despite having the same overall shape as other aircraft, is limited in where it can land, demanding runways of sufficient length for take-off and landing; in addition to appropriately sized handling facilities for the aircraft itself along with goods, luggage and passengers. If a blended wing-body aircraft was to enter the fleet, passenger loading and unloading and the implications for airport infrastructure layout would require a radical rethink; made all the more challenging as the changes would need to occur simultaneously at a global level.

Alternative Fuels

Perhaps the most promising solution on the horizon as far as the aviation industry is concerned is the potential for producing a kerosene-grade fuel from some form of biomass. Other alternatives, such as hydrogen or nuclear propulsion are problematic from both a perceived and real safety perspective. Furthermore, a hydrogen-powered aircraft would require a much larger fuel tank, involving complete aircraft redesign, where the problems faced regarding infrastructure and fleet renewal highlighted above would again come into play. Developing a fuel that can be dropped straight into existing designs is much more appealing. This could be as a proportion of existing fuel or the development of a fuel specifically aimed to replace fossil-fuel kerosene. Research and development into this is ongoing, but key sticking points remain. How sustainable is the source of biomass? What quantities are reliably available? What are the emissions associated with production and refinement? How do biofuels compete for land and water with food production and even conservation? In addition, demand for biomass-derived fuels has to compete for biomass demand from all other energy sectors, as well as potentially for its use as a chemical feedstock. The power sector, land-based transport, shipping sector and many other industries assume a percentage of bio-derived fuel contributing to their mitigation objectives, not to mention the use of biomass as an important source of carbon sequestration (through biomass combustion with carbon capture and storage). So a very real and pivotal question is what proportion of biomass production is it reasonable to assume the aviation sector can access? If some sectors have other opportunities for decarbonisation, and the aviation sector does not, is there an argument for such a fuel to be reserved for the aviation industry? How this situation could emerge politically is difficult to imagine, but is clearly an area for further debate and exploration.

CO_2 GROWTH FROM AVIATION

Comparing and contrasting recent and potential ongoing aviation and accompanying CO_2 emissions growth with postulated 2°C pathways demonstrates the significant gap between what is needed in terms of decarbonisation and the direction of travel of aviation-related CO_2 emissions. Considering first domestic aviation, Fig. 4 presents the trend in CO_2 by region. The dominance of domestic aviation in the United States is clear, but so too is how the proportion of total CO_2 from domestic flying within other nations is rising as their domestic aviation expands.

Fig. 5 has a similar time series but for international aviation. Whilst there was clearly a dip due to the recent economic downturn, in general growth rates are high, with signs that they may actually increase rather than decrease. Note the significant dip in non-OECD Europe in the late 1980s marking the collapse of the Soviet economy – another region with potential for a significant resurgence in growth.

Fig. 6 compares the growth trend in whole-economy CO_2 with that from domestic and international aviation, both as discrete entities and combined. It further illustrates how aviation has, despite only having developed

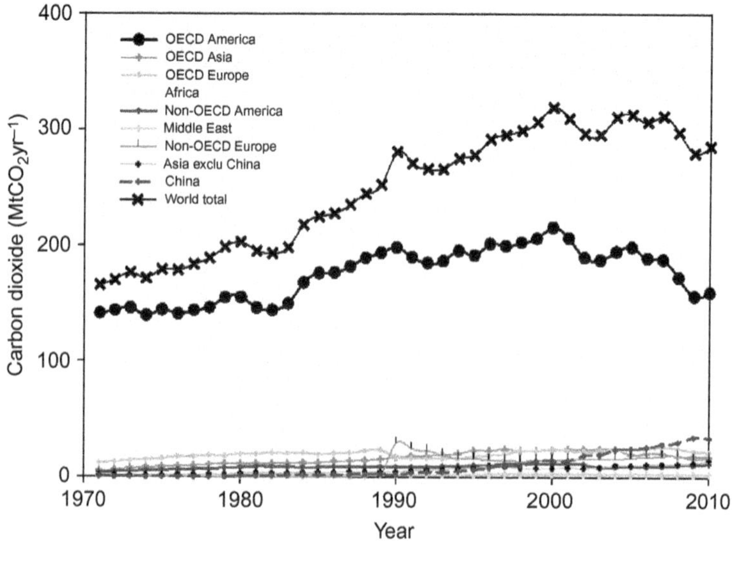

Fig. 4. Time Series of Domestic Aviation CO_2 Based on International Energy Agency Statistics.

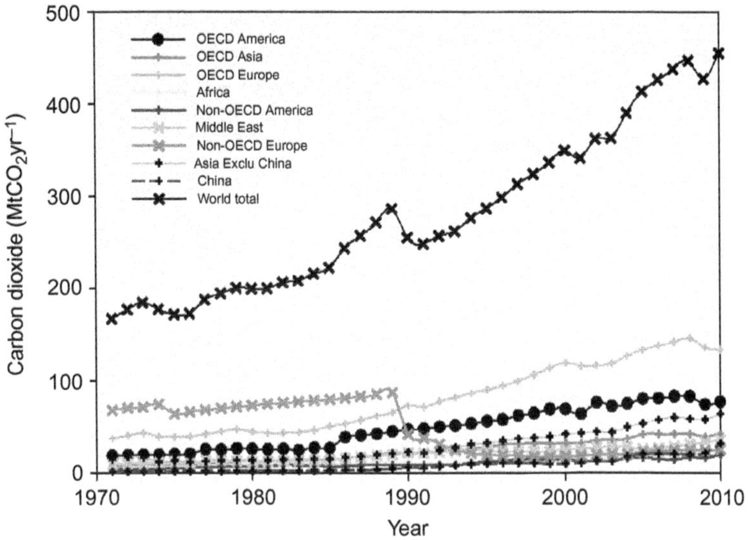

Fig. 5. Time Series of International Aviation CO_2 Based on International Energy Agency Statistics.

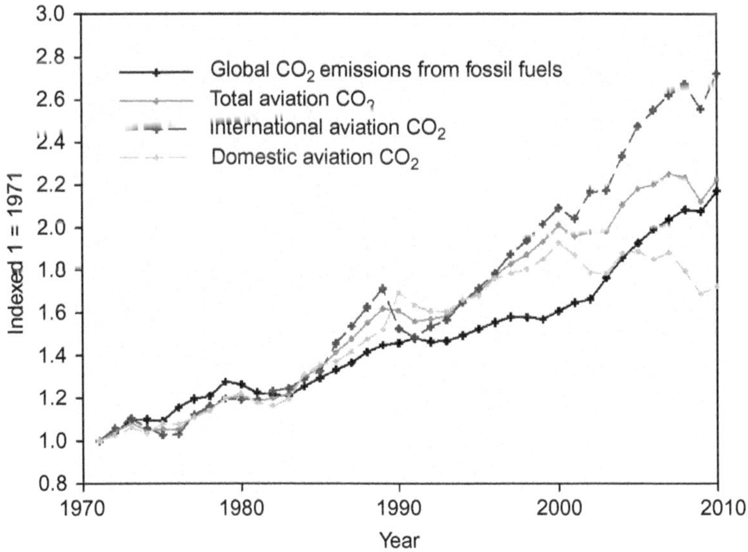

Fig. 6. Time Series of Data Indexed to 1971 for Comparing the Rate of Growth of Global Fossil-Fuel Produced CO_2 with that from Aviation.

significantly across a few nations, had emissions growth that outstrips total global CO_2 growth rates (from fossil fuel). Certainly from the mid-1980s until around 2008, aviation CO_2 tracked well above the trend for the aggregate of other sectors, with the growth in international aviation CO_2 specifically, even more pronounced. The growth rate of emissions from domestic aviation dropped below that of total fossil CO_2 after 2001 – coinciding with the events of September 11 and the subsequent impact on the popularity of domestic flying within the United States. International aviation was also impacted, though not as significantly and with a much faster return to previous growth rates. Fig. 6 also demonstrates how the economic downturn is playing out across both domestic and international aviation. Given a larger proportion of flying relates to leisure travel, this dip in growth in what many consider to be a luxury activity is, perhaps, to be expected.

CAN AVIATION EMISSIONS FIT WITH CLIMATE CHANGE TARGETS?

To explore the relevance of ongoing growth in CO_2 emissions from the aviation sector for climate change objectives, two simple scenarios are developed. These two scenarios illustrate how a continuation of growth in the CO_2 from aviation can, even with relatively conservative assumptions, lead to the sector dominating *total-economy* CO_2 emissions. No probabilities have been ascribed to the two scenarios, but both are plausible representations based on the issues discussed in the earlier sections. The scenarios both adopt a range of assumptions, as described below:

Scenario 1: Recent regional trends (2005–2010) in both domestic and international aviation continue for a further 40 years. Under such a scenario, CO_2 from the United States domestic aviation industry declines significantly, as demonstrated in Fig. 7(a), whilst the growing trend in China and Asia soon results in them becoming the largest emitters of domestic aviation CO_2. Note also that extrapolating such a trend, assumes continuation of the somewhat depressed growth brought on by the recent recession in industrialised nations. Between 2000 and 2010, the CO_2 growth rate for China's domestic flights averaged at 9% per annum and international at 12% per year. Such ongoing compound growth rates result in very high levels of absolute CO_2 emissions from China by 2050; though their per-capita aviation emissions, from domestic flights, is still below that of the United States 2000–2010 annual average.

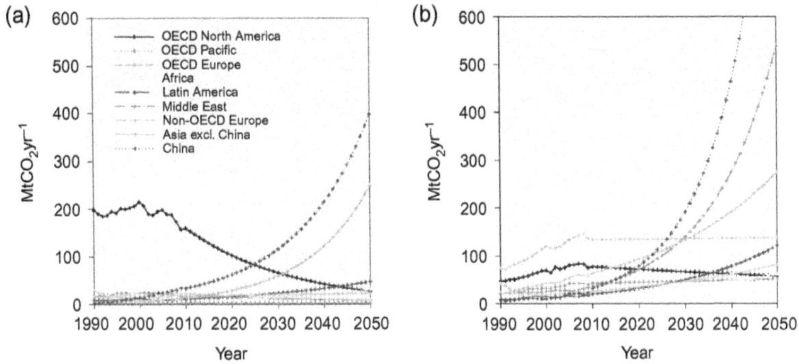

Fig. 7. Historical Domestic (a) and International (b) Aviation CO_2 Up Until 2010 Split by Region Followed by Scenario 1 Assumptions between 2010 and 2050.

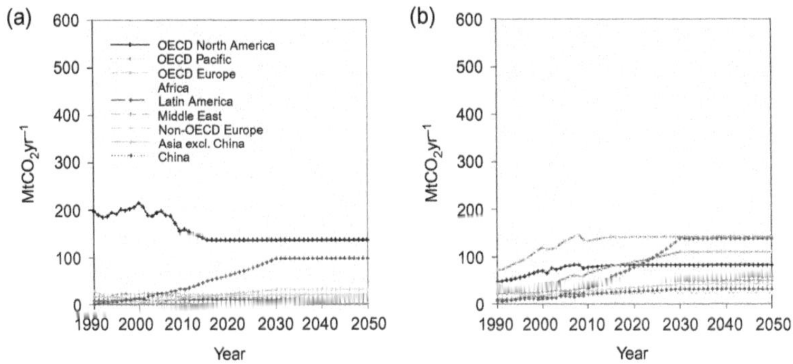

Fig. 8. Historical Domestic (a) and International (b) Aviation CO_2 Up Until 2010 Split by Region Followed by Scenario 2 Assumptions between 2010 and 2050.

Scenario 2: Recent regional trends (2000–2010) continue until 2015. Between 2015 and 2030 the growth rate is halved, and then post 2030 CO_2 emissions plateau in all regions. This results in much lower emissions of CO_2 by 2050 from the aviation sector (Fig. 8).

Fig. 9 combines the domestic and international aviation scenarios for comparison. Clearly if recent rates of growth continue post-2030, high levels of emissions will result. These scenarios are purely illustrative, but it is worth comparing them with the range of industry scenarios available.

The ICAO (2009) presents scenarios developed to explore the impact that a range of technological and growth assumptions could have on

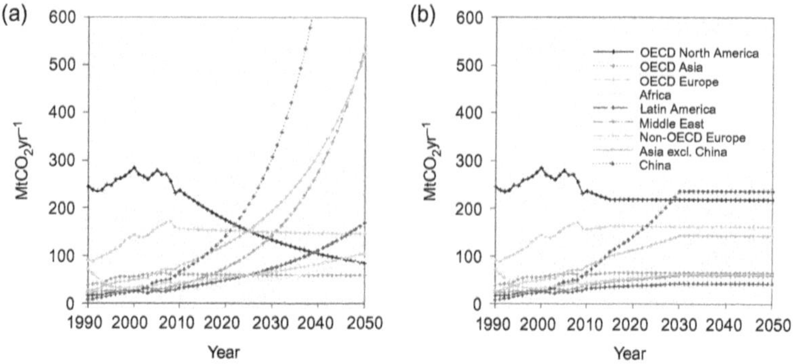

Fig. 9. Historical Total Aviation CO_2 Up Until 2010 Split by Region then Scenario 1 (a) and Scenario 2 (b) Assumptions between 2010 and 2050.

future global CO_2 emissions. Their scenarios range from emissions of 4,500 $MtCO_2$ by 2050 for a 'do nothing' scenario to 2,300 $MtCO_2$ for a very high technology scenario with moderate growth. This compares with a range of 1,000 $MtCO_2$–3,200 $MtCO_2$ in the two scenarios presented above. The low growth in emissions illustrated in scenario 2 is not currently under consideration by ICAO. They also state that the most likely range is between 2,300 and 2,800 $MtCO_2$. A review by Gudmundsson and Anger (2012) considered the full range of scenarios available in the literature, from which they estimated the mean scenario figure for 2050 to be 2,471 $MtCO_2$, with a mean minimum value of 1,444 $MtCO_2$ and mean higher value of 3,122 $MtCO_2$.

Comparing the two simple future scenarios developed here and the range of postulated global emissions of CO_2 discussed in the literature with global carbon targets serves to highlight the importance of tackling rising CO_2 from the aviation industry. Fig. 10 demonstrates that by 2050, total CO_2 emissions from fossil fuel combustion and industry, need to be around 3,900–5,400 $MtCO_2$ for a 50:50 chance of avoiding 2°C. If the aviation sector's emissions grow to between the ICAO 2050 levels of 1,444 $MtCO_2$ and 3,122 $MtCO_2$, then at a global scale, it will be responsible for a much greater share of future CO_2 emissions. However, under the ICAO scenarios, even by 2050, the prevalence of flying in some nations is below that evident in the United States or EU today.

The cumulative emissions framing of the climate change problem serves to remind policy and decision makers that reducing emission in the near term is essential to avoid the very high rates of emission reduction required

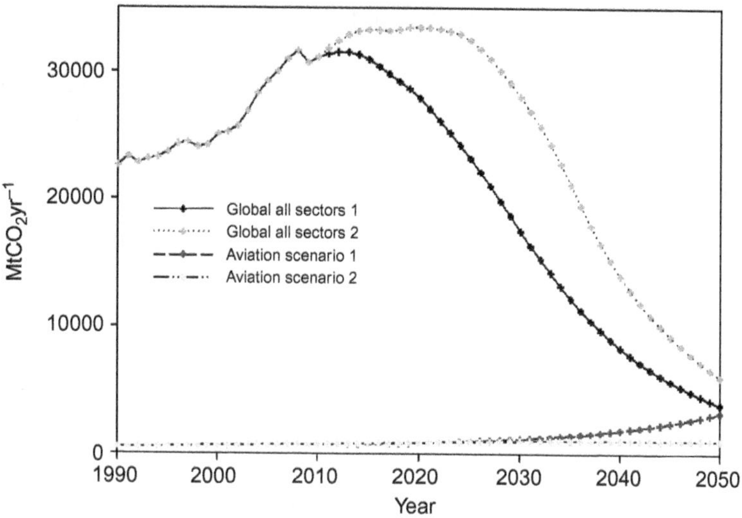

Fig. 10. Comparison of the Two Aviation Scenarios with the Global Emission Reduction Pathways Presented in Fig. 1.

in later years. Other sectors, particularly within transportation, are likely to struggle to meet the scale of emission reduction required from them. If just one industry, such as the aviation industry, does not reduce emissions to the same extent as others, or worse, allows emissions to continue to grow, then this places even more pressure on other sectors to decarbonise at higher and potentially infeasible rates. Or, alternatively, the climate objectives of G8 nations will be missed. Technological and operational advances can reduce the rate of CO_2 growth from aviation at present, but they are not going to deliver the required emission reductions. For instance, if the aviation sector set targets for emissions in line with what is required of all sectors on aggregate, globally its emissions would need to be reduced by some 75% in absolute terms from 1990 levels by around 2050 (Anderson & Bows, 2012). Arguably, the very high growth rates from this industry compared with others, if taken into consideration between 1990 and 2010, requires that aviation actually does more than the average, because it is the cumulative emissions that matter. Therefore, until technological and operational solutions can decarbonise air travel at rates well beyond the pace of expansion in the industry, policy measures must place a constraint on growth; for example, through delaying plans for increased airport capacity.

CONCLUSION

The policy debate around tackling aviation emissions now receives significantly more attention from academia, government, NGOs and the aviation industry than it did just a decade ago. Nevertheless, there remains an absence of any explicit regulatory, economic or voluntary measure yet to impact even the rate of growth in aviation-related CO_2.

The industry is regularly referred to as mature, and indeed if domestic aviation is analysed in isolation, CO_2 emissions have recently plateaued (Fig. 11). However, domestic aviation remains dominated by the United States, a highly industrialised and wealthy nation with a large landmass. The United States' mature aviation sector, September 11, SARS and the economic downturn have all aligned to produce a fall in US domestic aviation CO_2, with a subsequent major impact on the global rate of change. By contrast, China, with its similarly large landmass, has a rapidly growing domestic aviation sector set to influence significantly global domestic aviation emissions – potentially overriding CO_2 reductions from the more mature US aviation sector.

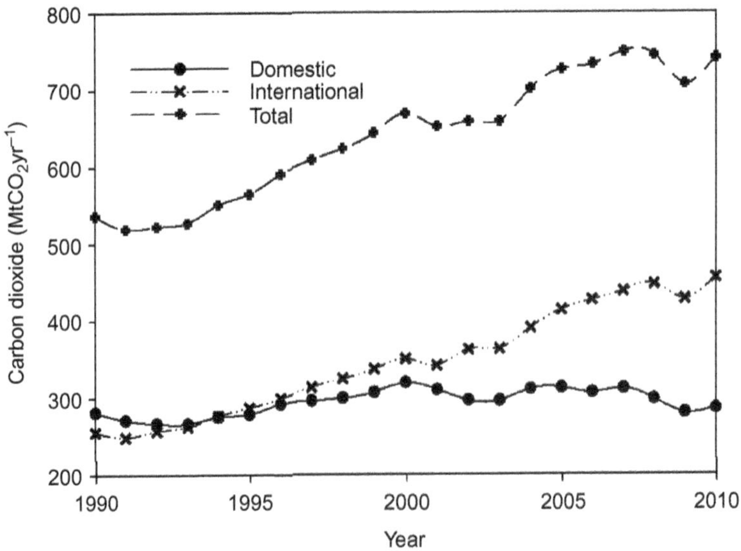

Fig. 11. Historical Domestic, International and Combined (Total) Aviation CO_2 from 1990 to 2010 from International Energy Agency Statistics.

International aviation, on the other hand, has not plateaued (Fig. 11). Apart from the recent economic downturn, growth rates are being upheld, with international links between emerging markets and established economies an important driver of growth. There is huge potential for ongoing growth in international aviation with associated increases in CO_2. While there are a suite of operational and technical opportunities for reducing the carbon intensity of aviation, the very high passenger growth rates mean they can only ever offset marginally the rate of CO_2 growth; they fall far short of what's necessary to cut emissions in absolute terms.

The clash between growing the aviation sector whilst simultaneously reducing emissions in line with international climate change commitments is emblematic of the wider global malaise towards evidence-based mitigation. Meeting the high-profile political commitment to stay below a 2°C rise in global mean surface temperature requires an urgent and rapid curtailment of global CO_2 emissions. On average, emissions need to be cut by around 75% by 2050 (from 1990 levels), even for a 50:50 chance of staying within the 2°C threshold. Yet aviation's CO_2 emissions are expected to grow by around 170–480% over the same period, and they could feasibly be higher still. Growth of this magnitude is entirely at odds with stated global climate ambition, even in a world where aviation is permitted some emissions-leeway. Given the severity of scale of the challenge to avoid 2°C, for any single sector to cut its emissions below that of the average, or even allow emissions to grow, demands still deeper cuts of other sectors. One suggested solution to this dilemma is the use of a carbon price – where the market rather than policymakers dictate which sector undertakes more or less mitigation. However, the level of carbon price necessary to deliver emissions-reduction commensurate with a 50:50 chance of avoiding 2°C would be economically destabilising and socially unacceptable.

Ultimately, if the international community is serious about its commitments on climate change, it needs to begin to debate some of the deeper moral and ethical arguments underpinning an effective climate response. Policymakers and wider society need to question whether supporting the ongoing expansion of an industry with few technological options for decarbonisation and which is largely used for leisure is a reasonable way to gamble with the future.

NOTE

1. http://ec.europa.eu/clima/policies/transport/aviation/index_en.htm

REFERENCES

Anderson, K., & Bows, A. (2008). Reframing the climate change challenge in light of post-2000 emission trends. *Philosophical Transactions A*, *366*(1882), 3863–3882.

Anderson, K., & Bows, A. (2011). Beyond 'dangerous' climate change: Emission scenarios for a new world. *Philosophical Transactions of the Royal Society A: Mathematical, Physical and Engineering Sciences*, *369*(1934), 20–44.

Anderson, K., & Bows, A. (2012). Executing a Scharnow turn: Reconciling shipping emissions with international commitments on climate change. *Carbon Management*, *3*(6), 615–628.

Anderson, K. L., Mander, S., Bows, A., Shackley, S., Agnolucci, P., & Ekins, P. (2008). The Tyndall decarbonisation scenarios – Part II: Scenarios for a 60% CO_2 reduction in the UK. *Energy Policy*, *36*(10), 3764–3773.

Bows, A., & Anderson, K. L. (2007). Policy clash: Can projected aviation growth be reconciled with the UK Government's 60% carbon-reduction target? *Transport Policy*, *14*(2), pp. 103–110.

Bows, A., & Anderson, K. (2008). *A bottom-up analysis of including aviation within the EU's Emissions Trading Scheme*. Working Paper No. 126.

Bows, A., Anderson, K., & Mander, S. (2009). Aviation in turbulent times. *Technology Analysis & Strategic Management*, *21*(1), 17–37.

Bows, A., Anderson, K., & Upham, P. (2008). *Aviation and climate change: Lessons for European policy*. London: Routledge, Taylor & Francis.

Department for Transport. (2003). *The future of air transport – White Paper*. London: The Stationery Office.

Gudmundsson, S. V., & Anger, A. (2012). Global carbon dioxide emissions scenarios for aviation derived from IPCC storylines: A meta-analysis. *Transportation Research Part D: Transport and Environment*, *17*(1), 61–65.

ICAO. (2009). *Report from GIACC/4*. Retrieved from http://www.icao.int/environmental-protection/GIACC/Giacc-4/CENV_GIACC4_IP1_IP2%20IP3.pdf

Meinshausen, M., Meinshausen, N., Hare, W., Raper, S. C. B., Frieler, K., Knutti, R., … Allen, M. R. (2009). Greenhouse-gas emission targets for limiting global warming to 2°C. *Nature*, *458*(7242), 1158–1162.

Scheelhaase, J. D., & Grimme, W. G. (2007). Emissions trading for international aviation – An estimation of the economic impact on selected European airlines. *Journal of Air Transport Management*, *13*(5), 253–263.

United Nations Framework Convention on Climate Change. (1997). *The Kyoto Protocol*. United Nations Framework Convention on Climate Change.

PART II
CHALLENGES

CHAPTER 4

ENVIRONMENTAL TECHNOLOGY AND THE FUTURE OF FLIGHT

Lucy Budd and Thomas Budd

ABSTRACT

Purpose – *To examine the role of new aeronautical technologies in improving commercial aviation's environmental performance.*

Methodology/approach – *Reviews the environmental improvements that may be conferred through the adoption of alternative aviation fuels and new airframe, engine and navigation technologies.*

Findings – *Although aeronautical technologies have evolved considerably since the earliest days of powered flight, the aviation industry is now reaching a point of diminishing returns as growing global consumer demand for air transport outstrips incremental improvements in environmental efficiency. The chapter describes some of the technological interventions that are being pursued to improve aviation's environmental performance and discusses the extent to which these innovations will help to deliver a more sustainable aviation industry.*

Keywords: Aviation; environmental impacts; future technologies

Sustainable Aviation Futures
Transport and Sustainability, Volume 4, 87–107

ISSN: 2044-9941/doi:10.1108/S2044-9941(2013)0000004004

INTRODUCTION

Commercial aviation is one of the most important components of the global economy yet also one of the most contentious. In a little over 100 years between the Wright brothers' first flights in December 1903 and today, the rapid development and subsequent ready-availability of safe, reliable and relatively cheap access to air travel worldwide has transformed the mobility patterns, employment prospects and consumption practices of millions of people on earth and global society's appetite for, and socio-economic reliance on, flying shows no signs of abating. According to the Air Transport Action Group (ATAG), 2.8 billion passengers and almost 48 million tonnes of airfreight (worth some US$5.3 trillion) were flown around the world on 26.7 million commercial flights in 2011 (ATAG, 2012). It is estimated that 56.6 million people worldwide are employed by the sector and commercial aviation's economic impact is thought to be in the region of US$2.2 trillion (equivalent to approximately 3.5% of the world's GDP). Indeed, commercial aviation's economic impact is such that if the sector were a country, it would be ranked 19th in the world by GDP (*ibid.*).

For those who can afford the price of an air ticket, the world's commercial airline network allows personal and professional relationships to be conducted at a distance and across multiple time zones. It enables tourists and vacationers to experience foreign countries, climates, and cultures and permits business travellers, students and migrants to rapidly access new commercial, educational, and entrepreneurial opportunities overseas. It allows politicians, diplomats, business leaders, academics and doctors to exchange knowledge and rapidly respond to natural disasters and humanitarian emergencies anywhere in the world. The coevolution of intricate and integrated just-in-time networks of air and surface logistics also enable high-value consumer goods, perishable foodstuffs and pharmaceuticals to be routinely and rapidly transported around the world from their place of production to their site of consumption. Although only comprising 0.5% of total trade volume, airfreight accounts for approximately 35% of all global trade by value (ATAG, 2012).

While the immediate socio-economic benefits of commercial aviation are difficult to refute, the routine aeromobility of tens of thousands of commercial aircraft and the combustion of millions of barrels of jet fuel a year generate a range of negative externality effects, the likely impacts of which are only now starting to be appreciated. Although some of aviation's deleterious environmental effects, including aircraft noise and atmospheric emissions, were recognised as early as the 1950s and have become synonymous with the aviation–environment debate, issues relating to commercial aviation's impact on human health, its role in the rapid international dissemination of agricultural pests and infectious diseases, and its contribution to anthropogenic climate change, have only more recently become the subject of sustained academic debate.

Although significant steps have been taken to reduce aircraft noise and improve environmental performance in recent years through the development and introduction of progressively more aerodynamic and fuel-efficient airframes and engines together with enhanced air traffic management procedures, commercial aviation is now rapidly reaching a point of diminished returns as global increases in the number of flights outstrips the incremental environmental efficiency gains afforded by the introduction of new technology. While some industry commentators and pro-aviation lobbying groups opine that continued technological innovations in the fields of material sciences, aerodynamics, precision area navigation and propulsion will provide viable and cost-effective solutions to aviation's environmental challenges, environmentalists and many airport communities are more sceptical. Indeed, for the sector's most strident critics nothing short of a fundamental step change in aeronautical technology and/or the complete abolition of flying would improve the sector's environmental performance.

In recognition of the increasingly urgent sustainability and public relations challenges the world's commercial aviation sector faces, this chapter critically appraises the role of selected technological innovations and interventions that have been proposed to mitigate some of aviation's principal environmental effects and assesses the extent to which they may individually and/or collectively help to improve aviation's environmental performance. Our focus here is very much on technologies that may reduce the environmental effect of aircraft in flight. The potential roles of new regulatory regimes, global emissions targets, demand management measures and aviation corporate social responsibility agendas are discussed elsewhere in this volume.

The chapter begins with a brief description of the scope of the contemporary aviation sustainability challenge. This is followed by a section that charts the historical evolution of aeronautical technology from the origins of heavier-than-air powered flight at the beginning of the twentieth century to the present day to explain the sector's contemporary reliance on carbon-based fuels. The third section describes and critically evaluates a range of technological proposals that have been advanced as a way to help improve aviation's environmental performance. The chapter concludes by suggesting that while any future environmental improvements and fuel efficiency gains are welcome, they must deliver tangible environmental benefits over existing technology that not only offset but ideally mitigate future increases in the number of flights. Given the apparent absence of quick technological fixes that will confer the carbon and emissions savings atmospheric and earth systems scientists consider necessary to avoid potentially catastrophic climate change (see Bows-Larkin & Anderson, 2013), a range of new demand management measures, that would undoubtedly be politically unpopular and socio-economically problematic to devise and implement in a globally equitable manner, may ultimately also need to be considered.

FROM GATE TO GATE: AVIATION'S ENVIRONMENTAL IMPACT EXAMINED

It is a well-known fact that aircraft noise and atmospheric emissions from aircraft engines respectively degrade the local acoustic environment and air quality around airports and contribute to anthropogenic climate change. In order to provide sufficient thrust to accelerate a passenger aircraft which, in the case of Airbus's A380 'Super Jumbo',[1] may weigh up to 560 tonnes, from a standing start into the air, climb it to an assigned cruising altitude of 35,000 ft or higher, and safely keep it there for the planned duration of the flight, the world's commercial aircraft fleet collectively consumes approximately 5,270,000 barrels of energy-rich and relatively volatile kerosene-based jet fuel a day (Chevron Aviation, 2006; EIA, 2012). The current generation of high-bypass turbofan engines that power contemporary wide-bodied aircraft can each produce around 80,000 lbs (over 36,300 kg) of thrust and, at take-off power, draw in over a ton of air through the front fan every two seconds (Snow, 2000). The act of drawing in cold air at the front of the engine, compressing a proportion of it, adding fuel to this compressed air, and igniting that fuel in the central combustion chamber, produces a constant exothermic reaction. The hot exhaust gases that are produced are directed out of the rear of the engine where they mix with the cold air that has bypassed the central core. The act of expelling hot exhaust gases and mixing them with cold air creates additional thrust and ensures an aircraft's wings can generate sufficient lift to overcome the combined effects of gravity and aerodynamic drag.

However, while jet fuel is an undeniably useful energy-rich power source, it produces a number of pollutants, including oxides of carbon, nitrogen and sulphur, methane, water vapour, particulates (soot) and non-methane volatile organic compounds (NMVOCs) when it is burnt. Depending on the altitude and latitude at which they are released, these pollutants can perturb the global climate and cause local air quality around airports to deteriorate. Near the ground, elevated concentrations of nitrous and sulphurous oxides degrade local air quality and have been implicated in a range of human respiratory and cardiovascular complaints. At higher altitudes during the cruise, carbon dioxide, water vapour and methane contribute to the radiative forcing of the atmosphere and are implicated in anthropogenic climate change.

While the white contrails that are produced by aircraft flying through saturated air at cruising altitude are arguably the most visible and familiar manifestation of aviation's environmental impact, the sustainability challenges the sector faces permeate every stage of the service delivery chain. Processes of airframe and engine manufacturing, maintenance and disposal are all highly energy intensive. They often involve the use of toxic chemicals and rare earth minerals and create complex man-made composites that can be problematic to reclaim, recycle and/or dispose of safely. The construction of airport passenger terminals, cargo sheds, runways, taxiways, maintenance areas and associated airside infrastructures all require the use of substantial quantities of raw materials (all of which have to be quarried and/or refined, and transported to the construction site). Building works inevitably disturb local habitats. They may also displace (as well as intrude upon) local residential communities and can require significant land take, often of greenfield sites. As a consequence, the construction and expansion of airports worldwide has proved controversial almost since the earliest days of powered human flight and many development schemes have been met with vociferous local opposition and public protest (see Bröer, 2013; Halpern, 2013; Knippenberger, 2013).

While anti airport expansion and anti-noise campaigns are a familiar manifestation of public anxiety about the environmental impact of aviation in many countries around the world, airport construction is only one aspect of air transportation that imposes an environmental burden. Once an airport is operational, routine turnaround and aircraft maintenance activities on the airfield collectively demand the use and/or disposal of toxic fuels, lubricants, hydraulic fluids, de-icing and anti-icing compounds, human waste, and catering refuse. Crash simulations and routine emergency exercises by airport rescue and fire fighting services necessitate the burning of kerosene and the use of toxic fire-retardant compounds while strict wildlife management policies – from regulating the height of trees under the final approach paths to runways, to bird displacement activities and the culling of problematic animal and bird species – are practised to protect the safety of aircraft and their occupants. All these activities, while designed to reduce the inherent risks associated with air travel to an acceptable level, also inevitably affect the local environment and local biodiversity around airports.

Routine operations in the passenger cabin of aircraft also generate a significant environmental footprint. In-flight meals demand the procurement, preparation, storage and carriage of fresh, culturally appropriate and appealing foodstuffs, the provision of disposable serving equipment and suitable temperature storage and reheating facilities prior to consumption. Used meal trays may contain co-mingled waste food, plastic, film, foil, paper, can and ceramic waste, much of which was, until recently, merely collected in refuse sacks and sent to landfill. Passenger utility and entertainment services, from aircraft lavatories (it has been estimated that one in-flight flush uses two litres of aviation fuel), to potable water supplies, in-flight entertainment systems, and the provision of miscellaneous items such as in-flight magazines and blankets require airlines to burn additional fuel to generate both the extra electrical energy that is needed to power the on board systems and the extra thrust that is required to keep the (now heavier) aircraft in the air. If the aircraft has arrived from an overseas country in which certain infectious diseases are endemic, the cabin may also have to be disinsected using powerful pesticides and insecticides to prevent the importation of infectious pathogens and protect human health.

On the ground, aircraft turnarounds and passenger terminals require significant quantities of electrical energy, gas, and potable and non-potable water. Waste disposal systems, capable of processing everything from human sewage and retail waste to obsolete, but still radioactive, x-ray and baggage screening equipment, must be provided. Telecommunications, security, air conditioning, fire detection, and heating systems are all now prerequisites of modern airport operations. Restaurants and retail outlets need regular supplies of (often chilled) perishable products, while staff, passengers, contractors and visitors may need to access the site 24-hours a day. As T Budd et al. (2011) have shown, cost, comfort and convenience dictate that most surface access journeys to and from passenger airports worldwide are conducted by private cars and taxis. The dominance of private vehicles in the surface mode split results in a significant deterioration in local air quality as well as delays and congestion on airport access roads.

Despite aviation's wide-ranging environmental effects, it is only within the last few decades that concerted attempts have been made to improve the sector's environmental performance. Many of the early efforts were driven by the need to improve fuel efficiency and lower costs during the oil crises of the 1970s whereas more recent initiatives have emerged in response to growing consumer and regulatory pressure for aviation to reduce its environmental impact and act in a more socially and environmentally responsible manner. In the next section we review the development of aeronautical technology from the first heavier-than-air powered human flights in 1903 to the present.

AVIATION TECHNOLOGY – THE FIRST 100 YEARS

The world's first heavier-than-air powered human flight occurred on the morning of 17 December 1903 on the windswept sand dunes of Kill Devil Hills, near Kitty Hawk, North Carolina, USA. Although only airborne for 12 seconds and barely covering a distance of 120 ft, Orville Wright became the first aviator in the history of human aeronautical endeavour to take off, successfully pilot an aircraft, and land at a point equal in elevation to that from which he departed. The Wright brothers' canvas and wood biplane marked an important moment in the history of aircraft design. Unlike the experimental gliders that had preceded it, the Wrights had fitted a gasoline-powered 12 horsepower internal combustion engine to their machine. This engine powered a propeller that was angled to pull the aircraft through the air and generate sufficient airflow over the wings to overcome the increased drag and the weight of the aircraft. Although the engine was noisy, underpowered, unreliable and generated noxious emissions, gasoline proved to be an ideal fuel as it was energy intensive, relatively cheap, reasonably safe to store and handle, readily available and, crucially, didn't congeal in cold temperatures at high altitude. As a consequence, gasoline-fuelled engines became the propulsion mode of choice for early aviators.

The years leading up to the First World War were marked by the progressive application of scientific knowledge to aircraft design. The outbreak of War in 1914 stimulated rapid advances in aircraft design and manufacturing techniques. The conflict identified a number of unique roles that aircraft could perform and a new range of aircraft were designed for purposes of aerial reconnaissance, surveying, and bombing. The war helped to identify the most efficient and reliable designs and, as time went on, designers quickly standardised on single wing monoplanes that were powered by one or more petrol engines that drove a front mounted propeller. The development of more powerful and reliable piston engines in the 1930s enabled designers to build larger and heavier aircraft and metal became the material of choice for aircraft construction.

The start of the Second World War in 1939 again stimulated rapid advances in aircraft technology and resulted in dramatic new innovations in aerodynamics, propulsion, navigation, and communication systems. The advent of the gas turbine (or jet) engine by Sir Frank Whittle and colleagues in England revolutionised firstly military and, a few years later, commercial aviation. Fuelled by special aviation-grade kerosene derived from crude oil, jet engines conferred significant increases in power and performance for only a modest increase in weight and they enabled aircraft to fly further, faster, longer and higher than ever before. Military jet fighters, which could outperform and outmanoeuvre the earlier propeller-driven aircraft, were quickly developed and, soon after the War ended, jet age technology was being applied to a new generation of post-war passenger aircraft. British aircraft manufacturer de Havilland, responding to the UK Government's call for a new generation of long-haul jet-powered passenger transport, designed the world's first jet powered commercial aircraft, the Comet. The Comet first flew in 1949 and, following a rigorous programme of ground and flight testing, entered revenue passenger service with British Overseas Airways Corporation (BOAC), a forerunner to the present day British Airways, on the London to Johannesburg route in May 1952. Despite its early promise, the Comet's commercial success was irrevocably damaged by a series of fatal accidents that were caused by metal fatigue that had resulted from repeated cycles of cabin pressurisation. Nevertheless, the jet engine had proved it was suitable to power commercial aircraft and other aircraft were designed to take advantage of the increased speed, power and range jet engines afforded.

In the United States, major American aircraft manufacturers including Boeing and Douglas began designing jet aircraft of their own. Boeing's B707, which could seat over 100 passengers and which was powered by four jet engines, entered service with Pan American Airlines on their transatlantic routes in 1958. Although much faster and more comfortable than the aircraft they replaced, the first generation of jet aircraft were very noisy and had a prodigious thirst for fuel. As a result, people living near major commercial airports started to complain about the peculiar shrill squeal of jet engines that would rattle their windows and render normal conversation impossible when aircraft passed overhead. In addition to being noisy, the early jets were also very inefficient by today's standards and very polluting. Thick black trails of soot could often be seen in the sky behind departing aircraft and this led to growing public concern about the environmental impact of flying.

The late 1960s and 1970s saw the introduction of ever-larger wide-bodied passenger aircraft, including Boeing's famous B747 'Jumbo Jet'. These machines could seat between 250 and 500 passengers and perform flights lasting longer than eight hours. A series of oil crises in the early/mid-1970s increased the price of crude oil and aviation fuel to record levels. As fuel constituted one of an airline's biggest costs, aircraft manufacturers came under increasing pressure to design more fuel-efficient aircraft. Engine manufacturers, in particular, now had a new commercial imperative to develop more fuel-efficient products that were not only safe but also cheaper to operate. This led to innovations in turbine design, which included the use of stronger and lighter composite materials, the development of new lighter metal alloys and the introduction of higher-bypass turbofans that progressively reduced the amount of fuel that was required to deliver an equivalent unit of thrust.

Rather than all the air passing through the central core of the engine, scientists discovered that if they mixed cold air with the hot gases in the exhaust plume they could further increase the power (thrust) of the engine without increasing fuel burn. The resulting new 'high bypass' turbofan engines were not only more powerful and fuel efficient, they were also less noisy than the earlier designs. Although high-bypass engines conferred significant improvements in environmental performance, even the most modern fuel-efficient engines still burn a finite carbon-intensive fossil fuel and emit a range of harmful pollution species. Similarly, despite notable reductions in engine noise, the acoustic energy that is created by aircraft remains a highly contested moral and geopolitical issue and airlines, airports, aircraft manufacturers, regulators and governments are under increasing pressure to reduce aviation's environmental impact still further and a number of potential solutions to these challenges have been proposed. In the next section of this chapter we critically review the extent to which suggested innovations in aircraft design, materials, alternative fuels and more sophisticated navigation systems can help to reduce aviation's environmental impact.

THE PROSPECTS FOR NEW TECHNOLOGY

In recognition of the need to improve aviation's environmental performance, a number of purportedly 'transformative' new technologies have been proposed. Some of these concern the design, materials and construction of the airframe, others refer to potential changes to propulsion systems and alternative power sources for aircraft while others concern technologies that seek to improve ATC efficiency and lessen the environmental impact of routine airport operations. It is to these technologies that the chapter now turns.

Materials

Historically, innovations in aircraft design were driven by demands for increased speed, reliability and performance. Over the course of a few years, canvas and wood biplanes were replaced by metal monoplanes and primitive engines surpassed by more powerful piston designs. The use of high-grade steel and the development of stressed-skin aluminium alloys from the late 1930s onwards allowed aircraft designers to build ever-stronger and more complex designs (Whitford, 2000). Continued innovations in aircraft materials resulted in the introduction of progressively lighter yet stronger aluminium and titanium alloys that were resistant to corrosion, cracking and fatigue. As a result, these materials were the metals of choice for passenger aircraft during the 1960s and 1970s. By the 1980s, however, it had been discovered that other materials, including carbon-fibre composites, could be manufactured to exhibit the properties required for aircraft structures. These carbon-fibre composites are made from strong carbon fibres that are set in a chemically and mechanically protective matrix of epoxy resin (Whitford, 2000). Crucially, these carbon-fibre composites are both strong and up to 20% lighter than aluminium alloys. Carbon-fibres were first used in a limited capacity on Airbus's A300 aircraft in 1980 and are now used by most aircraft manufacturers in a variety of structures including the tailfin and fuselage. It is estimated that the Airbus A340 would have been 11,595kg heavier if carbon-fibre composites had not been used (*ibid.*). Clearly these weight reductions are significant. Boeing's twin engine B787 Dreamliner is the first commercial passenger aircraft in history to have a fuselage built from these lighter carbon-fibre composites while over 20% of the A380 is made from composite materials, including carbon-fibre reinforced plastic.

The development of new polymers, composites and memory metals that are stronger but lighter than conventional aluminium alloys will lead to continued reductions in weight and fuel burn. Already, another composite material, glass fibre reinforced aluminium, is reported to be 25% stronger than conventional airframe grade aluminium but around 20% lighter.

Nevertheless, despite the weight reductions these new materials confer, the environmental impact of the processes involved in their manufacture and disposal need to be taken into consideration when the environmental benefit of these new materials is assessed. What is certain, however, is that over time research and development into new aircraft materials will enable the construction of new, perhaps radically different, aircraft.

Aircraft Designs and Airframe Configurations

In addition to reducing the weight of aircraft by using new materials, improving the aerodynamic efficiency of the airframe by lessening drag is another key tenet of future aircraft design, and various technologies and innovations have been proposed and tested in recent years, with varying degrees of success. The basic configuration of most modern subsonic jet powered passenger aircraft, with their thin cylindrical fuselages, underwing mounted high-bypass turbofan engines and swept back metal wings, has not changed dramatically over the last 60 years. This has led some aircraft designers to speculate on whether this conventional design remains the most effective and environmentally efficient. Certainly, a number of alternative configurations, including blended and strut-braced wings, which focus on altering the shape of the wings to improve their lifting properties while reducing drag, have been proposed.

Unlike standard aircraft where the wings are attached to the fuselage, with blended wing designs the wing structures are incorporated into the main structure of the aircraft. This gives the aircraft a more bulbous shape that, when viewed from above, resembles a delta wing. Interestingly, while the idea of a blended wing was first tested in the 1920s the lack of suitably strong materials prevented its development. Within the last few years, a number of scientists have proposed revisiting the design as they believe it would be more aerodynamically efficient and quieter than existing aircraft. Critics, however, have questioned whether passengers would be prepared to travel long distances in an aircraft that had few (if any) windows. Other proposed radical new aircraft designs include 'flying wings'. These designs share many of the characteristics of blended-wing aircraft but typically lack a defined fuselage or vertical stabiliser. Advocates of flying wings believe the design will confer much more favourable lift-to-drag ratios compared with conventional aircraft. However, while such designs may provide improved aerodynamic and structural efficiency, the absence of traditional stabilisation surfaces on the leading and trailing edges of the wing and empennage means that they may be more unstable and therefore difficult to control. As a result, blended-wing and flying wing aircraft may never enter commercial revenue service.

Other designs which have been advocated include strut-braced wings. This configuration, in which supporting struts are added between the wings and the fuselage, was also tested and largely abandoned in the 1920s but it is now being revisited. Adding supporting struts between the wings and the fuselage enables the wings to be lighter and longer. This reduces weight and drag and helps to increase the lifting area of the wing. Researchers in the United States have calculated that adding a supporting strut from the belly of the aircraft to the wings could enable airframe designers to reduce the weight of the wing by two thirds without compromising its strength or its ability to generate lift (Daviss, 2007). This, they believe, could improve fuel efficiency by 25% (Daviss, 2007).

An altogether easier, yet still effective, strategy to improve the aerodynamic properties of aircraft currently in service involves the addition of raked wingtips or wingtip fences. These devices are installed at the very end of the wings to smooth the interface between the turbulent airflows above and below the wings to reduce wingtip vortices and lessen drag. Most new aircraft are equipped with wingtips as standard but many older airframes are being retrofitted with wingtips to lessen drag and improve fuel efficiency. The ability to retrofit existing aircraft is very beneficial for airlines given the long life cycle of commercial aircraft and the capital expense of purchasing new ones.

A further wing-based innovation that has the potential to improve environmental performance is the use of laminar flow control technology. In the context of aircraft design, laminar flow describes the layer of air that passes over the wings and fuselage of the aircraft providing lift and keeping it airborne. The smoother the layer of air, the less drag and the less fuel you burn. Laminar flow control works by detecting and sucking turbulent air into the aircraft through tiny holes in the airframe. This has the effect of retaining a smooth layer of air close to the surface. It is estimated that the use of laminar flow control can reduce drag on the wings by up to 20% and result in a 10% fuel saving. However, research into laminar flow control largely ceased in the 1990s when fuel prices dropped and the cost of the equipment would have outweighed any advantages in terms of reduced fuel burn. Other ideas have included combining laminar flow control with a specially modified fuselage to cut engine thrust by up to 60% in the cruise and result in fuel savings of 20% or, more radically, employing military shape shifting designs that physically alter the shape of the aircraft according to flight stage to maximise efficiency (Daviss, 2007).

Engines

As well as changing both the materials that are used to build aircraft and the design of the airframes themselves, it may also be possible to significantly improve aviation's environmental performance by redesigning aircraft engines. Although very small incremental efficiency gains may be achieved with existing technology, others have proposed that a new generation of open rotors or propfan engines may represent the best opportunity to improve environmental performance. The proposed open rotor or propfan engines use a modified jet engine to drive specially shaped propellers that create less drag. Though slower and noisier than conventional jets, open rotor engines use less fuel and could achieve fuel savings of up to 30%. However, safety concerns about the effects of a blade failure on an open rotor engine may take time to resolve (on conventional engines, blade failures are contained by the cowling that surrounds the fan – open rotors do not have this protective barrier and so could, theoretically, strike the aircraft's fuselage if they broke free following a bird strike or a structural failure).

Power Sources

In addition to proposing alternative designs for engines, attention has also focused on the fuels that are used to power them. Concerted efforts are being made worldwide to develop fuels that are not only more sustainable to produce but which also do not emit as many environmentally damaging pollutants when burnt. One of the main challenges associated with developing alternative fuels for aviation is that gasoline and (later) kerosene-based fuels exhibit the qualities required of a fuel. They have also been used since the early days of aviation which means that not only is there a significant body of knowledge about the fuels but also that a substantial supply chain and system of fixed infrastructure has evolved to support their use. The embedded nature and historical inertia of jet fuel in modern aviation represents a major barrier to the search for alternatives as any new fuel must be a direct substitute that can be 'dropped in' to the existing fuel supply chain without recourse to expensive and time-consuming modifications. Nevertheless, aviation's reliance on jet fuel leaves airlines vulnerable to sudden changes in fuel price and the introduction of increasingly stringent environmental targets, emissions trading schemes and growing consumer concern about the environmental effects of flying, are now driving developments into alternatives.

One of the most promising developments relates to liquid biofuels. Unlike conventional jet fuel which is refined from crude oil, biofuels are created by chemically processing biomass (plant starches and sugars) to create a liquid energy source. While biofuels (particularly ethanol and plant oils) have been used to power road transport vehicles since the 1970s, it is only within the last couple of years that new blends that are suitable for use in aircraft have been developed. Aviation biofuels that are currently undergoing flight testing with major airlines worldwide have been produced from a range of feedstocks and biomass including jatropha, coconuts, algae, domestic refuse, woodchips and carmellia (an inedible green shrub).

Over 50 major airlines, including KLM, Thomsonfly, Virgin Atlantic, United and Air New Zealand, have performed test flights using different types and blends of biofuel on a range of different aircraft types in both revenue and non-revenue services. Significantly, the biofuels were considered to be as good as conventional jet fuel as the test flights reported no loss of engine performance. However, while biofuels are being promoted as a 'green' alternative to conventional jet fuel, a number of barriers to their widespread use remain. Currently, we are unable to produce enough biomass to replace conventional jet fuel. This has led to concerns that land will be used for biofuel crops rather than for food production which would have the effect of pushing up world food prices and potentially driving more people into food poverty. Other concerns relate to the high research and development costs of biofuel (and thus their relative expense versus jet fuel), uncertainties about the accounting procedures, the true life-cycle emissions savings of the fuels and issues relating to fuel consistency.

Although biofuels are being strongly advocated by some sections of the aviation community as a potential solution to aviation's environmental impact in the short to medium term, other possible power sources, including solar energy and hydrogen fuel cells, are also being explored. In the summer of 2010, a single-seat experimental solar-powered aircraft, the Solar Impulse, successfully completed a 24-hour test flight. The aircraft was fitted with propeller-driven electric engines that were powered by solar energy generated by the 12,000 photovoltaic cells on the upper surface of the wings. Other experimental flights using solar power and/or batteries have indicated that while they could power small (single or dual seat) aircraft they do not supply enough electrical energy to power larger aircraft and their inferior energy-to-unit mass ratio remains problematic. Indeed it has been estimated that batteries only produce 2% of the energy that is obtained from the same mass of petrol.

While the potential for using solar power and electricity to power aircraft is very limited at present, another potential power source, hydrogen fuel cells, have also been proposed. It is suggested that, following suitable refinement, hydrogen fuel cells could be used to drive electric motors that would power aircraft. However, the application of hydrogen fuel cells to aircraft is immature and while the use of such cells would eliminate pollution at point of use, the processes of producing the hydrogen and manufacturing the fuel cells both require energy and generate atmospheric pollution.

Enhanced Air Traffic Control Procedures

In addition to proposing and developing new aeronautical hardware, such as airframes, engines and alternative fuels, to reduce aviation's environmental impact, new technology, in the form of more sophisticated computer software and processing capabilities, is also enabling the more effective use of airspace. Increasingly advanced air traffic control procedures that use precision satellite navigation and multilateration radar enable aircraft to fly more efficient trajectories and operate more environmentally friendly arrival and departure routes. Many airports have refined their existing air traffic control procedures to facilitate more environmentally efficient continuous climb departures (CCDs) and continuous descent approaches (CDAs). CCDs and CDAs enable aircraft to continuously climb up to, and descend from, their cruising altitudes without being held at intermediate altitudes. The elimination of old 'step up' and 'step down' climbs and descents enables pilots to fly their aircraft in a more aerodynamic configuration for longer. This reduces drag and avoids continuous adjustments being made to engine thrust settings, both of which collectively lower emissions.

Once established in the cruise, the increased use of user defined trajectories (in which airlines and flight crew request the most fuel-efficient altitudes, headings and routings based on aircraft type, aircraft weight and en route weather conditions), better aircraft sequencing at airports and experiments to create 'perfect flights' all lead to a reduction in emissions from individual aircraft. Individual air navigation service providers (ANSPs) are at the forefront of developments to improve environmental efficiency and reduce the volume of emissions each flight generates. The UK's ANSP, NATS, has committed to reducing air traffic related carbon dioxide emissions by an average of 10% per flight by 2020, from a 2006 baseline (NATS, 2012a). They estimate that approximately 2% of this reduction will come from operational improvements in their air traffic control centres, a further 2% from the use of CCDs and CDAs at airports and 6% from reconfiguring airspace structure and introducing new technology (*ibid.*). A further NATS initiative involves calculating the three-dimensional inefficiency of the flights they control. Using a specially designed environmental metric, called the 3Di score, NATS can compare actual flight trajectories with an optimal or airline preferred trajectory that minimises carbon dioxide emissions (NATS, 2012b).

In addition to individual ANSPs improving the environmental performance of flights operating within their airspace, there is significant potential to improve airspace coordination internationally. European airspace, in particular, is a patchwork of fragmented sectors and control zones that were originally drawn up along sovereign territorial lines. This historical legacy means that European airspace is not optimised for environmental efficiency and aircraft often have to fly circuitous routes to avoid congested areas, to prevent overflying certain nations and to avoid the most expensive areas of airspace. Attempts to harmonise the existing European airspace network through the Single European Skies Initiative is being delayed by a lack of international consensus.

Airport Operations

In addition to improving the environmental performance of aircraft in the air, technology is also being used on the ground to make airport operations more environmentally efficient. Increasing the use of energy from renewable sources is one way in which airport operators are improving their environmental performance. Lighting, heating, cooling, servicing and ventilating large passenger terminals can be extremely energy intensive and expensive and so airports are developing new systems that can produce reliable and affordable sustainable energy and lower energy costs. Many airports have installed biomass boilers, worked to increase the amount of natural light and ventilation, and, in some cases, installed wind turbines to generate electrical energy and boreholes to exploit sources of geothermal energy. The use of solar panels to convert sunlight into electrical energy has already been implemented at a number of airports in North America and Australia and there is scope for using similar technology at Middle Eastern and (some) European airports. Airport tenant companies, including airlines and ground handling companies, are installing solar panels on their head offices and administration buildings to reduce energy consumption and some are working towards making their building estate carbon neutral. Airports are also trialling the use of electrically powered vehicles, are encouraging shops and catering suppliers to source local seasonable produce, are harvesting rainwater to flush lavatories and are encouraging their staff and passengers to arrive at the airport by more environmentally efficient forms of transport.

Surface Access

Prompted by a growing awareness of the environmental impacts of airport users' entire 'door-to-door' journey, technology is also being used to reduce the environmental impacts of travel to and from airports. The environmental impacts associated with surface access travel typically relate to issues of noise and visual intrusion, local air pollution, carbon emissions, loss of habitat and biodiversity, as well as other local environmental degradation (Caves & Gosling, 1999; Graham, 2008; Johnson, 1997). While aircraft emissions have traditionally been the focus of debates about the environmental impacts of aviation, the role of surface access travel should not be overlooked. For example, it is estimated that 80% of local air pollution at Heathrow Airport is derived from surface access traffic and airside vehicles (Humphreys, Ison, Francis, & Aldridge, 2005).

Technological innovations have typically been targeted at reducing the share of private vehicle trips in favour of promoting public transport use. Advanced transportation systems, such as personal rapid transit (PRT) systems, have been successfully implemented at a number of airports worldwide. These systems typically consist of small, fully automated carriages or 'pods' that run on a guide-way, such as an elevated track, to and from the airport terminal utilising an 'on-demand' style service. These systems may be attractive options for airport operators, as their costs and service attributes compare favourably with alternatives such as shuttle buses or automated people movers, and the level of emissions associated with them is negligible (Gavin & Duncan, 2005; Muller, 2005). Examples of existing PRT systems include the 'ULTra' system in operation at Heathrow Airport, and the '2getthere' service at Amsterdam Schiphol. While to date such systems have only typically been used to connect remote locations on the airport site (such as car parks) to the terminal building, there is clear scope for their extension to include trips from a wider area, for example from downtown regions.

Other initiatives are designed to encourage and facilitate public transport use. Access (or lack of access) to travel information is identified as a key factor in decisions to travel by public transport, and as such a number of airports offer mobile applications (or 'apps') that convey real time travel information to airport users about transport service schedules and attributes, information about delays and route planning facilities. The increased demand for traveller information services has been driven to a significant degree by the rapid technological advances and the use of smartphone and tablet devices in recent years (Marshall Elizer, Hoskins Squier, Brydia, & Beaty, 2012).

Other Technologies

Other suggestions for lessening aviation's environmental impact include revisiting older aeronautical technologies, such as the airship, and accepting slower forms of aero mobile travel. In an increasingly globalised world that relies on fast, efficient and safe air travel, the scope for the widespread use of such technologies is perhaps limited. Alternatively, much has been made of the possibility for conducting virtual meetings via advanced teleconferencing and internet video/phone technologies, which, at least in theory, negate the need to travel at all. To date, research into their effects on air travel is inconclusive with some studies suggesting that the availability of virtual means of meeting reduce demand for air travel, whereas others have shown that, conversely, they promote it.

CONCLUSION

Technological innovation has played a key role throughout aviation's historical development. Since the first powered flight by the Wright brothers in 1903, through the two world wars, the jet age and eventually on to modern day aviation, technological advances in aircraft design and manufacturing, engine design and navigation systems have been rapid. While the evolution of aeronautical technology has created all sorts of new opportunities for increasing the speed and volume of global travel and trade, it has also imposed a number of significant social and environmental costs that the sector has not always been quick to address. In recent years, however, growing concern and awareness of the environmental impacts of the industry has prompted a shift towards developing technologies that reduce aircraft noise and improve the sector's environmental performance.

This has led to significant progress in the design and construction of airframes, propulsion systems and alternative power sources. There remain, however, numerous barriers to the widespread adoption of many of these technologies in a commercial aviation setting in the short to medium term. In particular, the long life cycle and high capital expense associated with purchasing new aircraft means that new initiatives may take many years to filter through, while existing supply chains and systems of fixed infrastructure remain a significant barrier to widespread uptake of more sustainable fuels, not to mention concerns relating to the designation of land for growing crops for biofuels when millions of people around the world remain in food poverty.

Arguably the most successful innovations are those that can be retrofitted to existing aircraft and/or 'dropped-in' to established supply chains and fixed infrastructure as they can typically be implemented much more quickly and cost effectively. The recent trend for retrofitting raked wingtips or wingtip fences on current aircraft is a good example of such practices. While such modifications are perhaps unlikely to single-handedly 'solve' aviation's environmental problem on their own, smaller incremental changes can yield significant overall benefits when combined with a number of improvements in other areas.

Similarly, progress in technologies for improving ATC efficiency and lessening the environmental impact of routine airport operations are likely to yield the greatest benefit when they are implemented as part of a wider programme of technological innovation. The increased use of precise satellite navigation and multilateration radar for ATC systems and the growing trend for airports to use renewable energy sources for powering terminal operations are testament to the progress being made. It is important that decision makers continue to address potential barriers and pitfalls for future improvements, for example the on-going need to tackle political barriers associated with the implementation of the Single European Skies Initiative.

While much of the focus on aviation and the environment has understandably focused on aircraft and airport operations, it is also important that decision makers do not lose sight of the environmental impacts of associated activities, such as surface access. Here, it is likely that technology will have a role to play by giving airport users the ability to make more informed sustainable travel decisions, rather than actually cutting emissions at source.

An examination of environmental technology and aviation highlights a number of important wider issues. In debates about the role of environmental technology and aviation, questions relating to environmental and technological interdependencies persist. Perhaps most notably, while aircraft can be made to be significantly quieter *or* less polluting, achieving both at the same time is a much harder prospect. This raises difficult questions about the prospect of prioritising reductions in aircraft emissions over reducing noise impacts, or vice versa. In essence, such debates boil down to striking a balance between the immediate annoyance and social problems resulting from aircraft noise, with longer term issues relating to air pollution and climate change which are harder to quantify and may not be felt for years or decades to come.

Finally, it is important that advances in technological innovations are not seen as a 'get out of jail free' card and used to justify unsustainable future expansion; emissions savings from environmental technologies are likely to be lost if there are simply many more aircrafts in the sky. It is therefore vital that technology is developed and applied in a sustainable fashion, and that such innovations complement, rather than contradict, wider policy or fiscal measures aimed at ensuring a more environmentally sustainable aviation future.

NOTE

1. See Airbus.com (2012).

REFERENCES

Airbus. (2012). *Airbus – Dimensions and key data, A380*. Retrieved from www.airbus.com. Accessed on 17 November 2012.

Air Transport Action Group. (2012). *Aviation. Benefits beyond borders*. Geneva: ATAG.

Bows-Larkin, A., & Anderson, K. (2013). Carbon budgets for aviation or gamble with our future? In L. Budd, S. Griggs, & D. Howarth (Eds.), *Sustainable aviation futures* (Vol. 4). Bingley, UK: Emerald Group Publishing Limited.

Budd, T., Ison, S., & Ryley, T. (2011). Airport surface access management: Issues and policies. *Journal of Airport Management*, *6*(1), 80–97.

Bröer, C. (2013). Sustainability and noise annoyance. In L. Budd, S. Griggs, & D. Howarth (Eds.), *Sustainable aviation futures* (Vol. 4). Bingley, UK: Emerald Group Publishing Limited.

Caves, R., & Gosling, G. (1999). *Strategic airport planning*. Oxford: Pergamon.

Chevron Global Aviation. (2006). *Aviation fuels technical review*. Houston, TX: Chevron.

Daviss, B. (2007). Green sky thinking. *New Scientist*, 24 February, pp. 32–38.

Gavin,W., & Duncan, R. (2005). *Personal rapid transit at airports: Physical, operational, and financial considerations*. Conference paper presented at the 84th annual meeting of the Transportation Research Board, Washington, DC.

Graham, A. (2008). *Managing airports: An international perspective* (3rd ed.). Oxford: Butterworth-Heinemann.

Halpern, C. (2013). Airport companies as silent partners: The complex interplay between public and private ownership (Vol. 4). In L. Budd, S. Griggs, & D. Howarth (Eds.), *Sustainable aviation futures*. Bingley, UK: Emerald Group Publishing Limited.

Humphreys, I., Ison, S., Francis, G., & Aldridge, K. (2005). UK airport surface access targets. *Journal of Air Transport Management*, *11*(2), 117–124.

Johnson, T. (1997). *Environmental issues*. Conference presentation at the 2nd Annual European Convention on the Development of Surface Access Links to Airports, Chartered Institute of Transport, London.

Knippenberger, U. (2013). The development of Frankfurt/Main airport: a traditional narrative of loss and gain. In L. Budd, S. Griggs, & D. Howarth (Eds.), *Sustainable aviation futures* (Vol. 4). Bingley, UK: Emerald Group Publishing Limited.

Marshall Elizer, R. Jr., Hoskins Squier, D., Brydia, R. E., & Beaty, C. P. (2012). *Guidebook for implementing intelligent transportation systems elements to improve airport traveler access information*. Airport Cooperative Research Programme (ACRP) Report 70, Transportation Research Board of the National Academies, Washington, DC.

Muller, P. J. (2005). *Personal rapid transit, an airport panacea?* Conference paper presented at the 84th Annual Meeting of the Transportation Research Board, Washington, DC.

NATS. (2012a). *Corporate responsibility report 2012*. Retrieved from http://www.nats.co.uk/wpcontentuploads/2012/07/corporateresponsibilityreport2012.pdf. Accessed on 20 November 2012.

NATS. (2012b). *3di infocard*. Retrieved from http://www.nats.co.uk/wp-content/uploads/2012/07/3di_infocard.pdf. Accessed on 20 November 2012.

Snow, J. (2000). Airliner propulsion. In P. Jarrett (Ed.), *Modern air transport: worldwide air transport from 1945 to the present* (pp. 53–66). London: Putnam.

US Energy Information Administration. (2012). *Jet fuel use statistics*. Retrieved from www.eia.gov. Accessed on 9 November 2012.

Whitford, R. (2000). Structures and materials. In P. Jarrett (Ed.), *Modern air transport: worldwide air transport from 1945 to the present* (pp. 67–80). London: Putnam.

CHAPTER 5

AVIATION AND THE EU EMISSIONS TRADING SYSTEM

Annela Anger-Kraavi and Jonathan Köhler

ABSTRACT

Purpose – *This chapter considers the application of climate mitigation policies to the aviation sector with reference to the inclusion of aviation in the EU Emissions Trading System (EU ETS). Assessments of the possible economic impacts of including aviation in the EU ETS are reviewed and an impact analysis using the macroeconometric E3ME model is conducted.*

Originality – *The aviation sector is a significant and rapidly increasing source of GHG emissions. Because international policy measures have not been agreed, the EU has incorporated aviation in the EU ETS. It is therefore important to consider the possible economic effects of the ETS on the aviation industry and the wider economy.*

Methodology/approach – *The paper describes the approach used by the EU to include aviation in the EU ETS. Assessments of economic impacts have been made, but have often been limited in their approach. The paper complements the existing literature by including an economic analysis using the E3ME macroeconometric model of the EU that covers 41 industrial sectors including aviation.*

Sustainable Aviation Futures
Transport and Sustainability, Volume 4, 109–130

ISSN: 2044-9941/doi:10.1108/S2044-9941(2013)0000004005

Findings – *Microeconomic and macroeconomic assessments show the economic impacts of including the aviation sector in the EU ETS are small. The negative impacts, if any, on EU GDP and the air transport sector's economic output are less than 0.1% and 1% respectively. Distortions in competition, both between countries and industrial sectors, are therefore likely to be small.*

Implications – *In the long term (beyond 2020), including aviation in the EU can be seen as a positive move. If and when aviation is fully included in the EU ETS, and when the cost impacts of GHG emissions through permit prices are made evident, it is anticipated that airlines will start monitoring and reducing their GHG emissions by investing in new, less carbon intensive technologies.*

Keywords: EU ETS; CO_2 emissions; economic impacts; air transport; emissions trading

INTRODUCTION

The EU is currently the world's largest provider of international air transport services (ICAOData, 2009; IEA, 2009a). EU airlines carry about a quarter of the world's passengers and about a quarter of the world's freight and about half of all these passengers and freight are carried by the airlines of three EU Member States: the United Kingdom, Germany and France.

The growth of the air transport industry has been accompanied by an increase in its environmental impacts, including climate change. Although carbon dioxide (CO_2) emissions from aviation comprised about 3% of global emissions in 2008, aviation CO_2 emissions in the EU alone increased by 110% between 1990 and 2008 (IEA, 2009a) to account for 20% of global aviation-derived CO_2 emissions. CO_2 is often considered to be the most important greenhouse gas (GHG) owing to its role in radiative forcing, but is not the only climate-impacting pollution species emitted by aircraft. Other emissions include NO_x, water vapour and particulates (soot). If GHG emissions from aviation continue to increase, the climate impacts are going to become more severe and extensive. It is therefore desirable to limit or mitigate the quantity of CO_2 and other GHG emissions that airlines can emit.

On the international policy side, aviation's growing impact on the global climate has provoked intense national and international debate on reducing

GHG emissions from aircraft. At present, only GHG emissions from domestic air transport operations are covered by the Kyoto Protocol to the United Nations Framework Convention on Climate Change (UNFCCC). International aviation emissions are left to ICAO (the International Civil Aviation Organisation) to regulate, and ICAO has not reached an agreement since 1997. The European Community therefore decided to take unilateral action and include CO_2 emissions from fuel burned by aircraft in its Community Emissions Trading System (EU ETS).

Quick and easy technology fixes for reducing carbon emissions from aircraft are scarce, although some operational and maintenance measures, such as reducing aircraft weight and washing aircraft engines more frequently, do reduce fuel burn and confer some environmental benefits (see for example Morris, Rowbotham, Angus, Mann, & Poll, 2009). New, less carbon-intensive aeronautical technologies, such as airships (Ghanmi & Sokri, 2010), open rotor propulsion and blended wing airframes might be available in the future, but not before 2050 (e.g. Köhler, 2012; see also Budd & Budd, 2013).

This chapter considers the application of climate mitigation policy to the aviation sector, in particular the inclusion of aviation in the EU Emissions Trading System (EU ETS) and the possible impacts on the sector. Firstly, we review the history of environmental policy in aviation. Then we describe the development of the EU ETS and discuss the contestation surrounding aviation's inclusion in it. Assessments of the possible economic impacts of including aviation in the EU ETS are reviewed. An impact analysis using the E3ME macroeconometric model of EU countries is then performed and the findings are discussed.

ENVIRONMENTAL POLICY IN AVIATION

Currently, the environmental performance of the aviation sector is regulated by national authorities, international agreements and voluntary arrangements. The UN body ICAO is, among other things, responsible for the environmental regulation of international commercial aviation. ICAO, through its Council's Committee on Aviation Environmental Protection (CAEP), primarily focuses on aircraft noise and the impact of aircraft engine emissions (ICAO, 2010a). All commercial aircraft are required to meet ICAO's strict engine certification standards on noise and pollution.[1] In addition, ICAO promotes the use of operational measures and develops guidance for its Member States on the application of measures aimed at reducing or limiting the environmental impact of aircraft engine emissions and mitigating aviation's climate change impact.

Other organisations from within the aviation sector have also set themselves voluntary targets to reduce their environmental impacts. For instance, members of the Advisory Council for Aeronautics Research in Europe (ACARE) agreed voluntary targets to improve the environmental performance of aircraft. The four defined goals, agreed for the time period 2002–2020, are (SRA, 2002):

1. To reduce fuel consumption and CO_2 emissions by 50% by developing more efficient aircraft and aircraft engines, improving air transport management and introducing low or zero carbon fuels;
2. To reduce perceived external noise by 50% by developing quieter aircraft and helicopters, developing low-noise flight profiles and introducing better community impact management schemes;
3. To reduce nitrous oxides (NO_x) by 80% by developing 'clean', that is low NO_x engines.
4. To make substantial progress in reducing the environmental impact of the manufacture, maintenance and disposal of aircraft and aircraft-related products.

In 2007, International Air Transport Association (IATA) member airlines adopted a fuel efficiency goal to achieve at least a 25% overall fuel efficiency improvement in their collective fleet by 2020 compared to 2005 (IATA, 2012). On 4 June 2007, the CEO of IATA became even more ambitious and called for a zero emissions future for the air transport industry (IATA, 2007). IATA (2010) now compiles a database (EBPDB, 2010), accessible to everyone, of voluntary measures and best practices that are undertaken by different stakeholders in the aviation sector.

One of the most important regulations governing fuel-use related emissions that affect the climate is Article 24 of the Chicago Convention (2006). This Article is implemented in bilateral agreements and, as commonly interpreted, states that no taxes or duties can be levied on aviation fuel. The argument against taxing fuel for aviation is that unilateral taxation would result in 'tankering', that is purchasing excess fuel in countries where no taxes are levied (RCEP, 2002). Article 24 is generally interpreted to mean there is no limitation on taxing fuel that is used for domestic flights. The European Commission's Directive 92/81/EEC (1992) exempts aviation fuels from excise duty for flights within EU Member States – only jet fuel used for private pleasure flying is taxed. These exemptions can be seen as subsidising the industry in violation of the 'polluter pays principle' (RCEP, 2002). These regulations are politically difficult to renegotiate (especially since there are some 4,000 bilateral agreements); therefore countries have to use national taxes. For example, the United Kingdom levies an Air Passenger Duty (APD)[2] that was designed to reduce the environmental distortions caused by the aforementioned exemptions (Mayor & Tol, 2007). The UK Government doubled the APD in 2007 and increased the rates again from December 2009 (HMRC, 2010).

The Kyoto Protocol to the United Nations Framework Convention on Climate Change (Kyoto Protocol, 1997) is an international and legally binding agreement to reduce greenhouse gas emissions worldwide. The Kyoto targets do not include aviation emissions from international flights. However, Article 2 (2) states that developed countries 'shall pursue limitation or reduction of emissions ... from aviation and marine bunker fuels, working through the International Civil Aviation Organisation and the International Maritime Organisation, respectively' (Kyoto Protocol, 1997). GHG emissions from fuel consumption in international aviation (and international shipping) are reported separately in national inventories as memo items under the category 'International Bunkers' (EEA, 2010). Hence, GHG emissions from domestic aviation are covered by the Kyoto Protocol and reported under the common reporting format as 'Domestic aviation'. Military and private (also called business or general) aviation are not included in either of these categories and are usually reported under the 'Other' category (IPCC, 2006).

The overall effect of these policy initiatives so far in reducing the growth of emissions from aviation has been limited. Despite a contribution of only about 3% to annual global CO_2 emissions, aviation is arguably the most environmentally unsustainable of the main modes of transport currently available (Chapman, 2007). This is due to the rapid expansion and estimated future growth of the sector resulting from globalisation, increasing GDP (incomes), liberalisation of the air transport market, and appearance of new business models that enable more people to travel by air than ever before. All this, combined with current aircraft technology that does not allow for quick and easy fixes to reduce CO_2 emissions and other climate impacts of aviation, has led to projections of substantial increases in emissions from air transport and has made policymakers consider climate policies for aviation. Integrating aviation into the EU ETS is the first policy measure to regulate aviation's CO_2 emissions at an international level (Anger & Köhler, 2010).

REVIEW OF THE EU ETS

GHG emissions trading is a relatively recent climate change policy intervention in Europe. The scheme, known as the European Union Emissions Trading System (EU ETS), came into force in 2005 and is currently in its third phase. International GHG emissions trading for the majority of developed countries (the Kyoto Protocol Annex B countries that have ratified the protocol) under the rules of the Kyoto Protocol started in 2008. Pre-EU ETS experience with emissions trading has mainly been in the United States, where, for example, permits to emit sulphur dioxide have been and are traded. It is generally considered that the US experience has been positive in reducing emissions at a reasonable cost (Ellerman, 2005).

The EU ETS is currently the world's largest emission trading system and the first emissions trading system that crosses country borders (for more on the political economy of emissions trading see MacKenzie, 2007). It is the centrepiece of current European climate change policy. The Emissions Trading Directive (Directive 2003/87/EC, 2003) was adopted by the European Parliament in June 2003 (for more on the history of developing the EU ETS, see Ellerman, Convery & de Perthuis 2010, pp. 9–31).

The EU ETS includes CO_2 emissions from energy intensive industries in the European Union (Directive 2003/87/EC, 2003) and it came into operation on 1 January 2005. The first phase (Phase 1) ran from 2005–2007 and the second (Phase 2) from 2008 to 2012. The third and current phase covers the years from 2013 to 2020 (SEC, 2008). The overall number of allowances in the EU ETS and annual allocation of tradable permits to participating companies are defined in advance of trading. The success of the EU ETS depends on its ability to establish a high enough and stable carbon price that gives an incentive for companies to undertake mitigation actions and invest in low(er) carbon technologies. The first phase failed to deliver a high enough carbon price mainly because of the generous allocation of permits in the EU market that caused the spot price of a carbon allowance to fall in May 2006. In May 2007, after the European Commission announced the verified emissions for the year 2006 and it was clear that the Phase 1 allowances were not transferrable into Phase 2,[3] the carbon price fell close to €0.

Despite all this, the first phase of the EU ETS was an important learning phase for creating the European carbon market and providing valuable carbon trading experience for companies (Ellerman & Joskow, 2008). The European carbon market has developed into a well-established commodity market that comprises a spot market as well as a derivatives market.

Currently the EU ETS covers about 50% of total EU carbon emissions. Following the experience with heavy industry, the EC expressed an interest in including the transport sector in the ETS. The aviation sector was the first transport sector to be included, from 2012 (Directive 2008/101/EC, 2009). Inclusion of two other transport sectors, marine bunkers and road transport, remain in the initial concept stage (European Commission, 2013).

For the first two trading periods each EU Member State was required to ensure that operators hold a GHG emissions permit and have a National Allocation Plan (NAP). Each NAP stated the total quantity of allowances (EUAs) that the country will receive for that period and the allocation to the operators of each installation. One allowance permits the emission of one tonne of CO_2 equivalent during a specified period (Directive 2003/87/EC, 2003). So far, most of the allowances have been allocated using 'grandfathering'. For the first trading period, 95% of allowances had to be distributed for free and 5% had to be auctioned. For the second period, the proportions were 90% and 10% respectively (*ibid.*). The free distribution of allowances could be seen as an incentive to support the adoption of the scheme since it causes less political opposition (Boemare & Quirion, 2002). The EU ETS allows the use of credits from Kyoto flexible mechanisms, Clean Development Mechanisms (CDM) and Joint Implementation (JI) projects,[4] up to a certain limit that is approved by EC (Directive 2004/101/EC, 2004 and Directive 2009/29/EC, 2009). For example, in the second phase of the EU ETS (i.e. 2008–2012), it was possible to use credits from CDM and JI projects up to about 13.4% (i.e. about 300 million CDM and JI credits) of allocated emissions (NAPs, 2008). Member States could also establish rules for new entrants, closures, early action, as well as for additional installations/activities that could be excluded from or included in the scheme (so called opt-ins and opt-outs) (Directive 2003/87/EC, 2003).

The EU ETS is growing out of its learning phase and maturing (Montagnoli & de Vries, 2010). The third phase of the EU ETS runs from 2013 to 2020. It has more centralised and harmonised rules – to avoid the mistakes made during the first two phases and to ensure more equitable treatment of trading companies (SEC, 2008). Phase 3 is designed to reduce emissions that are covered by the EU ETS by 21% from 2005 levels by 2020.

The first trading period of the EU ETS was meant to be a trial, and the first valuable lessons were drawn. The verified emissions of the first trading year (2005) of the first trading period were much lower (approximately 6% less) than the amount of allocated allowances (European Commission, 2008). When this was announced, the price of carbon dropped. Thereafter, the carbon market experienced high market volatility followed by gradually decreasing prices. The latter was caused by oversupply of carbon allowances and by the fact that most of the Member States (excluding France and Poland) did not foresee inter-temporal 'banking' between the first and second trading period (COM, 2006). However, intra-temporal banking and borrowing within the first phase of the EU ETS was permitted. Oversupply of allowances was driven by states' generosity in allocating allowances (that could interpreted as State Aid[5] (Johnston, 2006)), and possible hidden zero or negative marginal abatement costs inside industries (it is possible that if energy costs are a small proportion of overall costs, little attention will have been paid to energy efficiency). There may be simple operational measures or small investments that save energy and reduce operational costs, such that overall costs of operation are decreased. The latter is difficult to assess by governments because of the lack of relevant information, known as the principal-agent problem.

On the positive side, despite the collapse of the price of carbon, the carbon trading system was successfully established – and it works. Moreover, Ellerman and Buchner (2008) show that the phase one of the EU ETS saved up to 5% emissions in the EU ETS sectors. Bredin and Muckley (2010) explored the carbon price in the European carbon market and its drivers and found that in the second trading period the relationship between energy and carbon prices is getting closer to the ones theoretically predicted and the market is starting to show signs of efficiency. However, as of summer 2013, the price of emissions has again fallen and is around €4 per ton of CO_2.

AVIATION IN THE EU ETS

The European Parliament decided to include aviation in the EU ETS from 2012 (Directive 2008/101/EC, 2009). Aviation's inclusion was the EU's response to ICAO's delay in reaching an agreement on GHG reduction measures for aviation. This can be interpreted as the EU taking unilateral global leadership in tackling the problem and as a call for a global action to reduce the climate impacts of aviation. This view is supported by the fact that the EU is considering abandoning the inclusion of aviation in the EU ETS if equivalent international measures under UNFCCC or ICAO are adopted (*ibid.*). In fact, the EC proposed in 2012 that flights which operated into and out of Europe in 2010, 2011 and 2012 could be exempted to provide negotiation time for the ICAO General Assembly in autumn 2013, although the legislation continued to apply to all flights within and between the 30 European countries in the EU ETS (COM, 2012). The official statement declared that 'The Commission believes a global solution is within reach at the 2013 ICAO General Assembly' (COM, 2012). In its statement the Commission made clear that, should the meeting fail to make the necessary progress, the EU ETS legislation would be applied in full again to all flights to and from European airports. At the time of writing, this still has the status of an EC proposal and is not yet EU law.

The main points of the Directive 2008/101/EC (2009) are:

1. *Accountable entities:* All operators that operate scheduled or non-scheduled flights carrying passengers, freight and mail are included – unless their activities fall under the exclusion criteria discussed below. The inclusion of airlines is irrespective of nationality and business model. The decision is in line with the 'polluter pays' principle of international environmental law. Alternatives that have been discussed for accountable entities include airports, fuel suppliers, providers of air traffic management and aircraft manufacturers (Frontier Economics, 2006; SEC, 2006; Wit et al., 2002).
2. *Activities covered:* All flights that arrive at or depart from an airport situated in the territory of the EU are covered by the scheme. This means that both EU and non-EU operators are covered. In this decision the EC aligns with the Chicago Convention that requires equal treatment of all aircraft operators. However, the unilateral nature of the scheme contradicts the same requirement of the Convention, since it does not cover the airline operators that do not fly within the EU. Also, whether the flights within international and national airspace of third parties belong under EU jurisdiction can be challenged. In January 2010, the US Air Transport Association took the case to the English High Court, which in turn referred it to the European Court of Justice – the only body that can rule on EC directives (ATA, 2010).
3. *Activities excluded:* Annex I of the Directive lists the aviation activities that are not included in the EU ETS and the Commission Decision 2009/450/EC (2009) provides a detailed interpretation. The scheme will exclude state and military flights, flights under visual flight rules (VFR),[6] training flights, circular flights, flights performed for checking equipment, flights performed by aircraft with a certified maximum take-off weight of less than 5,700 kg, official missions from third countries and flights performed on routes within outermost regions. The

scheme specifies a *de minimis* rule that excludes commercial aircraft operators that perform less than 243 flights from EU airports in three consecutive four month periods or those who emit less than 10,000 tonnes of CO_2 per annum. These flights are omitted from the ETS because of the disproportionately high administrative costs associated with including them. However, general aviation operators do not qualify under the *de minimis* rule because they do not hold an air operator's certificate (AOC) under Part I of Annex 6 to the Chicago Convention and, therefore, need to comply.

4. *Timescale:* The first year for aviation in the EU ETS is 2012. As described above, international flights were excluded until 2014. The first full trading period of the EU ETS that includes aviation is the third trading period, starting from 2013 and likely to run until 2020.
5. *Greenhouse gases covered:* CO_2 is the only aviation-related GHG included in the EU ETS. No attempt will be made to cover non-CO_2 climate impacts of air transport. For these impacts (e.g. NO_x, contrails), additional policy instruments and interventions will need to be developed (COM, 2008).
6. *The number of allowances allocated:* Including air transport in the EU ETS will require allowance allocations at three different levels:
   (a) From 2012, the number of allowances available in the EU ETS (i.e. cap) will be increased by the number of allowances that will be allocated to the air transport sector (explained in point b).
   (b) Air transport's allocation will be based on historical emissions (i.e. on grandfathering). In 2012, the total number of carbon allowances allocated to aircraft operators will be set *at 97% of the total average annual GHGs emitted in 2004–2006 by these operators.* This allocation will then be lowered to 95% for 2013 and subsequent years of the trading period.
   (c) For the aircraft operators, a benchmarking approach will be used to allocate allowances. The benchmark equals the number of allowances to be allocated for free (from point a) divided by the tonne-kilometre (distance × payload) data from the benchmark year. Passenger aircraft can use the actual weight of a passenger and their luggage for calculations, or use a default value of 100 kg instead. The number of allowances allocated to an aircraft operator equals the benchmark multiplied by the tonne-kilometre data included in the application. The first benchmark period (reference year) will be 2010; thereafter, the calendar year ending 24 months before the start of the period to which the auction relates will be used.

7. *Allowance distribution methodology:* Allowance distribution will be harmonised across all EU Member States. *15% of the allowances will be auctioned and 85% granted for free to the aircraft operators.* The number of allowances auctioned by a Member State is proportional to its share of the total aviation emissions for all Member States for the reference year. The Commission takes a decision on the rules of auctioning 15 months before the start of each trading period.
8. *Making use of credits from Kyoto flexible mechanisms:* Certified Emissions Reductions (CERs) of CDM projects and Emission Reduction Units (ERUs) of JI projects from Kyoto flexible mechanisms will be used up to a harmonised limit that will be set by the EC for the third trading period. For the second trading period (year 2012 for airline operators), the limits that are set in the NAPs will be used – that is airlines are allowed to use CERs and ERUs up to about 15% of their EU ETS allocation in 2012. From 2013, the use of CERs is unclear.
9. *Opt-out and opt-in possibility:* An opt-out possibility is an option to exclude certain areas or activities from the EU ETS. The opt-out possibility is foreseen for third countries, overseas countries and territories, and ultra-peripheral regions. The directive allows for the exclusion of third country airlines that already face comparable measures. Such exclusions will be decided through international negotiations. An opt-in possibility is an option to include certain activities or entities with emissions under the limits stated in the directive.
10. *Nature of the aviation trading system:* The trading system will be open, with the only restriction being that the non-aviation operators cannot surrender aviation allowances. This is because the allowances that are issued for airlines under the EU ETS are not backed with the Kyoto allowances or included in the Kyoto targets. A closed trading system for the aviation operators only was also discussed as a possible option to avoid the problem with aviation allowances not being backed by the Kyoto units. A closed scheme was considered to be too costly for the aircraft operators and, therefore, politically infeasible (SEC, 2006).
11. *New entrants:* A special reserve of allowances (3% of total quantity of allowances) will be established for airlines entering the EU ETS during the trading period and for airlines that grow by an average of more than 18 % annually.

It should be stressed that including aviation in the EU ETS is a part of a comprehensive package of measures to tackle the climate change impact of aviation. The other measures proposed by the European Community include operational and technological measures (Directive 2008/101/EC, 2009).

POTENTIAL IMPACTS ON THE AVIATION SECTOR OF ITS INCLUSION IN THE EU ETS

Potential Impacts

The aviation sector's CO_2 emissions have been growing more than 3% per year (IEA, 2009a) despite rising kerosene prices (ICAOData, 2009) and the imposition of other costs (e.g. the Air Passenger Duty imposed in the United Kingdom in January 2001). This suggests that similar policies that result in increasing costs for the industry, such as the EU ETS, may have modest effects.

Fig. 1 provides a simplified schematic representation of the impacts of carbon price (either allowance price or tax on airline CO_2 emissions) through the demand-side reaction on airlines. The figure shows that the impact of a tax or a trading scheme depends, in general, on how sensitive consumers and airlines are to the price increase caused by the policy. The cost of carbon may be passed through to passengers through tickets price increases. Depending upon the elasticity of demand, this may decrease demand and decrease profits. Airlines therefore have an incentive to change their operations or invest in alternative technologies that reduce emissions and so reduce their costs. The figure is a simplified representation because the actual response can be much more complex due to external factors such as transaction costs, competition, fuel prices and possible route switching.

However, airlines can respond to the allowance costs, which can be interpreted as a surcharge on fuel costs, through supply-side reaction. In economic terms, the supply-side reaction of the air transport industry to a carbon tax or emissions trading system depends on the price elasticities of fuel demand. An increase in fuel costs can encourage airlines to search for measures that help to save fuel and/or switch to lower-carbon alternatives or biofuels. CO_2 emissions from biofuels that are part of the Earth's carbon cycle are not currently included in emissions trading systems or covered by a carbon tax (e.g. see Directive 2003/87/EC, 2003).

Estimation of Impacts

The economic impacts of the inclusion of the aviation sector in the EU ETS have been assessed using both microeconomic and macroeconomic methods. The literature on impacts (see Anger & Köhler, 2010 for a review) of including aviation in the EU ETS shows that inclusion of the aviation

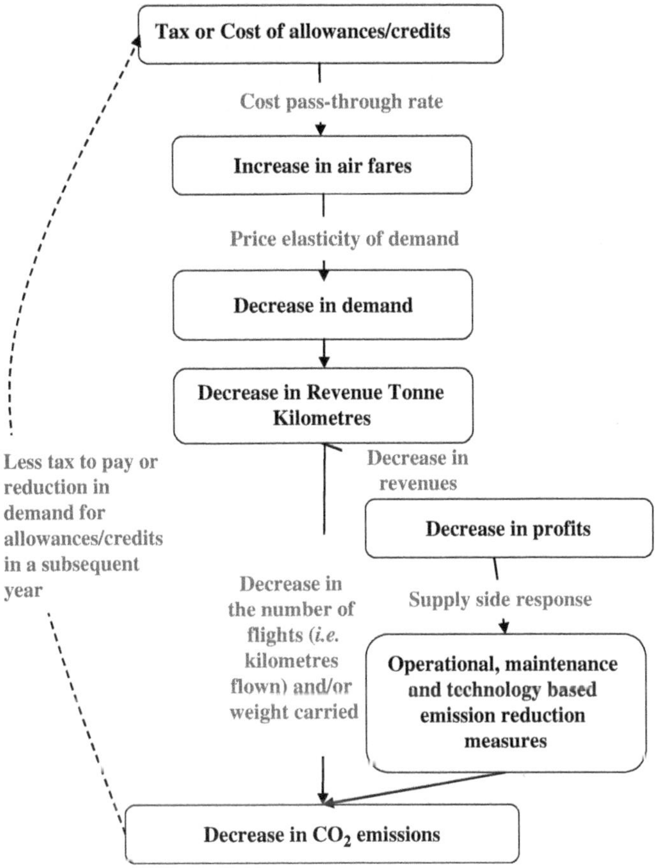

Fig. 1. Simplified Schematic Representation of the Air Transport Sector's Response to the Introduction of an Emissions Trading System or Carbon Tax.

sector will result in small reductions (a maximum of 3.3%) in growth rates of the airline industry and its emissions by 2020. The growth impact remains generally small for different open emissions trading system designs and exogenous allowance prices from €6 to €60 per tonne of CO2. The aviation sector is expected to purchase allowances from the other sectors covered by the EU ETS and use relatively lower-cost credits from the two Kyoto flexible mechanisms – CDM and JI.[7] The amount of allowances needed by aviation will be small compared with the size of the trading scheme, and therefore, aviation will not significantly affect carbon prices in the market. Airlines will pass the cost of purchased allowances, as well as the cost of freely allocated allowances (that bear opportunity costs), on to consumers. Their reaction is dependent on the price elasticity of demand, which is argued to be inelastic for business travellers and relatively elastic for leisure travellers. A study by Mason (2005) suggests that the actual price elasticity of demand for business travellers is higher than estimated in previous studies.

Empirical evidence from the air passenger duty imposed in the United Kingdom in January 2007 does not show significant reduction in passenger numbers (a maximum of 1.2%) and no change in CO_2 emissions (Mayor & Tol, 2007). These results are the same for both Low Cost Carriers (LCCs) and Full Service Carriers (FSCs). This and a report by the UK CAA (CAA, 2005) suggest that the current demand for LCCs might be much more inelastic than assumed in these studies. If this holds, then the impacts of increased costs caused by the purchase of carbon allowances will be even smaller than presented above. Also, the distortions in competitiveness between LCCs and FSCs will be smaller.

An analysis using a macroeconomic model is reported in Anger (2011), Anger (2010) and Anger and Köhler (2010). The Energy-Environment-Economy Model for Europe (E3ME) is a hybrid Post Keynesian macroeconomic dynamic simulation model. It is designed to assess short and medium run (up to 2030) GHG mitigation policies, including emissions trading systems (see, for example: Lutz and Meyer (2010), Köhler, Jin, and Barker (2008), SEC (2008) and Barker, Junankar, Pollitt, and Summerton (2007)). The model is a combination of time-series econometric relationships (estimations are based on data covering the period 1970–2006) and cross-section input-output (for the year 2000) relationships. Air transport is one of the 42 industrial sectors in the model. Hence, E3ME can simulate air transport in interaction with 41 other industrial sectors in a particular region (an EU Member State) and in a group of regions (the EU as a whole). The model also allows for the effects of changes in fuel use on emissions of the six Kyoto GHGs and other atmospheric pollutants.

Different scenarios of auctioning allowances and allowance of the inclusion of credits from the Kyoto Clean Development Mechanism were calculated. The main results are reported as percentage changes from the emissions trading *scenario REF* which has no aviation in it (Table 1). In all the scenarios the design of the EU ETS follows the EU regulations on emissions trading as closely as possible. The only major difference is that in the aviation emissions trading scenarios (*MIN, A, A1, B* and *B1*) the air transport sector has a diminishing allowance allocation from 2013 to 2020

Table 1. Impacts of Aviation Emissions Trading Scenarios (Percentage Differences from the *scenario REF* for the Year 2020).

Aviation Emissions Trading Scenario	Aviation Sector's Output	CO_2 Emissions from Aviation	EU GDP	EU CO_2 Emissions
MIN*(no auctioning)*	−0.576	−6.886	0.564	−1.062
A*(15% of auctioning)*	−0.552	−6.879	0.605	−1.051
B*(100% of auctioning)*	−0.491	−6.871	0.706	−1.036
A1*(15% of auctioning and* ***no*** *CDM credits)*	−0.535	−6.930	0.628	−1.058
B1*(100% of auctioning and* ***no*** *CDM credits)*	−0.474	−6.922	0.729	−1.045

Source: Anger (2011).

while in the current regulation the allowance allocation will remain at 97% level of the average aviation emissions of 2004–2006 for this period.

The EU ETS allowance price for the *scenario REF* was found to be €20.70 (at 2008 prices); all emissions trading scenarios with aviation (Table 1) had an allowance price of €28.75 for the trading period 2008–2020. This is likely to be the maximum increase in the allowance price due to inclusion of aviation, because the modelling assumes that the air transport sector needs to cover all CO_2 emissions emitted above the initial allowance allocation using allowances and CDM credits bought from the market, and the modelling does not utilise any no-regret options available to the aviation industry (emissions reduction options are discussed in Stratus, 2005; SEC, 2006; EBPDB, 2010; Morrell, 2009 and Morris et al., 2009). For instance, air transport's CO_2 emissions could be cut by reducing the amount of excess fuel and/or water, as well as baggage carried on board during a flight.

The policy assessment using E3ME shows that inclusion of the aviation sector as it stands in Directive 2008/101/EC(2009) will result in a maximum of 0.6% reduction in output of air transport by 2020, compared to the no-action reference scenario. The E3ME modelling (Anger, 2011) also shows larger reductions of 6.9% in CO_2 emitted by the air transport industry compared to other estimates in the literature. This is mainly a supply-side response to raised costs and investments from increased profits that appear because of the passing on of opportunity costs to consumers. The overall effects on EU's CO_2 emissions were rather small, showing 1.1% reduction. Anger (2011) also analysed impacts on real GDP in the EU in 2020. The E3ME modelling shows an increase of 0.6% in real GDP (in 2000 prices) in the EU for *scenario A* compared to the *scenario REF* in 2020. The positive effect on GDP is for the most part attributable to the increased (up 2.9% in 2020) economic activity in government sectors due to allocation of auctioning revenues to these sectors.[8] This finding does not take into account the impact of market shares shifting to non-EU carriers if they are allowed to avoid compliance. A comparison of impacts on the various Member States shows slightly larger impacts on old States than new States, although the results are mixed. This result comes from the larger share that the aviation sector makes up in old Member States. All auctioning revenues were used to increase Member States' government expenditures.

In the E3ME modelling, three scenarios were used to assess the impact of auctioning upon the industry and economic activity in the EU. *Scenarios A* and *B* (see Table 1) differ according to the levels of auctioning. *Scenario A* has a fixed auctioning level of 15% as in the current legislation for aviation emissions trading. *Scenario B* has 15% auctioning in 2012, then 20% in 2013, which will, thereafter, increase up to 100% in 2020. In addition to the *scenarios A* and *B*, a *scenario MIN* that includes no auctioning and in which all allowances are distributed at no cost, is examined. The different levels of auctioning have almost no impact on CO_2 emissions and output of the industry (Table 1). This is because the E3ME modelling assumes 100% cost pass-through to consumers. Nevertheless, the auctioning of allowances impacts aviation industry's profits. The revenues made by passing on 100% costs of free allowances is assumed to increase airlines profits and these do not increase the shares of profits invested, for example in R&D.

In this analysis, government spending is spread equally between defence, education, health and other sectors. For example, these activities are often intensive users of surface transport that is not covered by the EU ETS and increasing activities in these sectors tends to also increase CO_2 emissions. Therefore, *scenario MIN* (no auctioning) shows the highest decrease (1.06%) in EU level CO_2 emissions below *REF* in 2020. The impact of auctioning on EU GDP is positive for all scenarios showing a maximum of 0.71% increase above *REF* in *scenario B*. This is due to increased economic activity generated by auctioning revenues. If no revenue is generated via auctioning aviation allowances (*scenario MIN*), the GDP in the EU is increased by 0.56% compared with the *REF scenario*; this raise stems from the higher carbon price and hence increased recycling of auctioning revenues from other EU ETS industries.

Employment of CDM projects' credits for compliance purposes was also studied (Table 1), and it can be considered, while comparing *scenarios A* and *B* with CDM credits with *scenarios A1* and *B1* with no CDM credits that, despite a slight reduction in CO_2 emissions at the air transport industry and the EU level, there is some improvement in efficiency, that is a reduction in compliance costs for the air transport sector because CDM credits are generally cheaper than the EU allowances.

Overall, both microeconomic and macroeconomic analyses show that inclusion of aviation in the EU ETS is not expected to reduce demand growth or carbon emissions significantly and it is unlikely to encourage the uptake of new technologies in the long run. This is because aviation emissions are a small proportion of the overall emissions in the EU ETS and due to the relatively low carbon prices that are expected in the EU ETS after 2012. On the other hand, a closed trading system solely for aviation is very likely to have such effects, but is not likely to be politically feasible. However, in response to society's expectations, the EC proposal has drawn attention to the increasing climate change impact of aviation and this might encourage the sector to change its behaviour by using cost-effective options (shown in EBPDB, 2010 and Stratus, 2005) in the short to medium term.

SUMMARY AND RECENT POLICY DEVELOPMENTS

The concept of emissions trading is to use the market to implement emission reductions at the lowest cost to achieve a predefined emissions reduction target. Industries where emissions abatement is expensive 'fund' abatement in industries where it is cheaper. In effect, through engagement in the EU ETS, the aviation industry will 'pay' for emission reductions where mitigation costs are lower, improving the overall economic efficiency of emissions mitigation.

Inclusion of aviation in the EU ETS can be seen as the EU response to ICAO's delay in reaching an agreement on GHG reduction measures for aviation. This can be interpreted as the EU's unilateral global leadership in tackling the problem, and as a call for a global action to reduce the climate impacts of aviation. This view is supported by the fact that the EU has expressed a willingness to delay the implementation for international flights to wait for the outcome of the ICAO negotiations in 2013.

The economic impacts of including the aviation sector in the EU ETS are assessed as being small by both microeconomic and macroeconomic assessments. The negative impacts, if any, on EU GDP and air transport sectors' output are less than 0.1% and 1% respectively and therefore distortions in competition, between countries and industrial sectors, are likely to be small.

In the long term (beyond 2020), including aviation in the EU can be seen as a positive move. Once it is announced that aviation is to be included in the EU ETS, and the cost impacts of GHG emissions are made evident, companies in the aviation industry will start seeking to reduce their GHG emissions, for example, by investing in new, less carbon intensive technologies. At the aviation industry level, it is already possible to observe increased interest in new technologies and other GHG reduction measures.[9] The EU ETS for aviation has been subject to discussions at public and policymakers' level, domestically and internationally. This has drawn attention to aviation's GHG emissions and may result in similar actions by other countries and regions and even lead to a global climate change mitigation policy for air transport. Although many countries still do not support mitigation policies for aviation, the prospect of aviation emissions increasing while other sectors that are included in the EU ETS mitigate their emissions suggests that the policy and social pressure on aviation to take an active role in GHG mitigation will increase. If ICAO does not come to an agreement, there is still the possibility that aviation will have to take an active part in the EU ETS. If the assessments prove to be correct and there are no major impacts on the industry, the case for stronger mitigation policy will be enhanced. A probable near-term outcome is that the industry will develop the use of climate neutral biofuels as they offer the most rapid technical possibility of emissions reductions (see Köhler, forthcoming).

NOTES

1. http://www.icao.int/env/Standards.htm
2. APD is a boarding tax and not an environmental tax but, by increasing ticket prices, it acts as an environmental tax and has the potential of reducing CO_2 emissions and other environmental impacts from air travel.
3. While there was no formal ban on banking between Phase 1 and Phase 2 of the EU ETS, the EC announced that banked allowances would be deducted from Phase 2 National Allocation Plans (NAPs) and therefore Member States decided not to allow banking between Phase 1 and Phase 2 (COM, 2006).
4. CDMs and JIs are GHG reduction projects carried out in developing countries (Non-Annex 1 countries of UNFCCC) and economies in transition (Annex 1 countries of UNFCCC), respectively (Kyoto Protocol, 1997).
5. State Aid is an intervention (in any form) granted by an EU Member State to a firm that distorts or threatens to distort competition and is likely to affect trade between Member States. Based on http://ec.europa.eu/competition/state_aid/overview/what_is_state_aid.html

6. VFR flights differ from commercial traffic that operates under Instrument Flight Rules (IFR) under the jurisdiction of air traffic control.
7. For the second trading period, the maximum usage of these credits was 22% of allocated allowances (see European Commission's Decisions for NAPs for the second trading period – http://ec.europa.eu/environment/climat/2nd_phase_ep.htm).
8. The study uses a reference case that does not allow for the effects of the global economic recession.
9. For example, see GreenAir Online – http://www.greenaironline.com/index.php

REFERENCES

Anger, A. (2010). Including aviation in the EU ETS: Impacts on the industry, CO_2 emissions and macroeconomic activity in the EU. *Journal of Air Transport Management, 16*, 100–105.

Anger, A. (2011). *Emissions trading for regulating climate change impacts of aviation: A case study of the European Union Emissions Trading System.* PhD thesis. University of Cambridge.

Anger, A., & Köhler, J. (2010). Including aviation emissions in the EU ETS: Much ado about nothing? *Transport Policy*, *17*(1), 38–46.

ATA. (2010, May 27). English High Court permits ATA legal challenge to EU Emissions Trading System to proceed. ATA press release, Washington, DC. Retrieved from http://www.airlines.org/News/Releases/Pages/news_5-27-10.aspx. Accessed on June 24, 2010.

Barker, T., Junankar, S., Pollitt, H., & Summerton, P. (2007). Carbon leakage from unilateral environmental tax reforms in Europe, 1995–2005. *Energy Policy*, *35*, 6281–6292.

Boemare, C., & Quirion, P. (2002). Implementing greenhouse gas trading in Europe: Lessons from economic literature and international experiences. *Ecological Economics*, *43*, 213–230.

Bredin D., & Muckley, C. (2010). *An analysis of the EU Emission Trading Scheme.* Working Paper 201003. Geary Institute, University College Dublin. Retrieved from http://www.ucd.ie/geary/static/publications/workingpapers/gearywp201003.pdf. Accessed on October 26, 2010.

Budd, L., & Budd, T. (2013). Environmental technology and the future of flight. In L. Budd, S. Griggs, & D. Howarth (Eds.), *Sustainable aviation futures* (Vol. 4). Transport and Sustainability. Bingley, UK: Emerald Group Publishing Limited.

CAA. (2005). Demand for outbound leisure air travel and its key drivers. Civil Aviation Authority, London. Retrieved from http://www.caa.co.uk/docs/5/ERG_Elasticity_Study.pdf. Accessed on June 19, 2007.

Chapman, L. (2007). Transport and climate change: A review. *Journal of Transport Geography, 15*(5), 354–387.

COM. (2006, November 29). Communication from the Commission to the Council and to the European Parliament on the assessment of national allocation plans for the allocation of greenhouse gas emission allowances in the second period of the EU Emissions Trading System, COM(2006) 725 final, 2006. Retrieved from http://eur-lex.europa.eu/LexUriServ/site/en/com/2006/com2006_0725en01.pdf. Accessed on July 18, 2011.

COM. (2008). Communication from the Commission to the European Parliament, the Council, the Economic and Social Committee and the Committee of the Regions – Strategy for the internalisation of external costs. Retrieved from http://eur-lex.europa.eu/

LexUriServ/LexUriServ.do?uri=CELEX:52008DC0435:EN:HTML:NOT. Accessed on January 24, 2009.

COM. (2012). Proposal for a DECISION OF THE EUROPEAN PARLIAMENT AND OF THE COUNCIL derogating temporarily from Directive 2003/87/EC of the European Parliament and of the Council establishing a scheme for greenhouse gas emission allowance trading within the Community. COM(2012) 697, European Commission, Strasbourg. Retrieved from http://ec.europa.eu/clima/policies/transport/aviation/index_en.htm. Accessed on March 20, 2013.

EBPDB. (2010). Aviation environmental best practice database, IATA, 2010. Retrieved from http://www.iata.org/whatwedo/environment. Accessed on October 12, 2010.

EEA. (2010). Annual European Community greenhouse gas inventory 1990–2008 and inventory report 2010. Technical Report 6/2010. European Environmental Agency, Retrieved from http://www.eea.europa.eu/publications/european-union-greenhouse-gas-inventory-2010. Accessed on June 27, 2008.

Ellerman, D. A. (2005). US experience with emissions trading: Lessons for CO_2 emissions trading. In B. Hansjürgens (Ed.), *Emissions Trading for Climate Policy: US and European Perspectives* (pp.78–95). Cambridge: Cambridge University Press.

Ellerman, A. D., & Buchner, B. K. (2008). Over-allocation or abatement? A preliminary analysis of the EU ETS based on the 2005–06 emissions data. *Environmental Resource Economics, 41*, 267–287.

Ellerman, A. D., Convery, F. J., & de Perthuis, C. (2010). *Pricing carbon: The European union emissions trading system*. Cambridge: Cambridge University Press.

Ellerman, A. D., & Joskow, P. L. (2008). The European Union's Emissions Trading System in perspective: A Report prepared for the Pew Center on Global Climate Change. May 2008. Retrieved from http://www.pewclimate.org/docUploads/EU-ETS-In-Perspective-Report.pdf. Accessed on August 17, 2009.

European Commission. (2008, May 23). Emissions trading: 2007 verified emissions from EU ETS businesses. Press release, Brussels. Retrieved from http://europa.eu/rapid/pressReleasesAction.do?reference=IP/08/787. Accessed on October 17, 2008.

European Commission. (2013). Proposal for a regulation of the European parliament and of the council on the monitoring, reporting and verification of carbon dioxide emissions from maritime transport and amending Regulation (EU) No. 525/2013, COM(2013) 480 final, 2013/0224 (COD), Brussels.

Frontier Economics. (2006, March). Economic consideration of extending the EU ETS to include aviation: A report prepared for the European Low Fares Airline Association (ELFAA). Retrieved from http://www.elfaa.com/documents/FrontierEconomicsreportforELFAA-Economicconsideration_005.pdf. Accessed on May 14, 2008.

Ghanmi, A., & Sokri, A. (2010). Airships for military logistics heavy lift: A performance assessment for Northern operation applications. Ottawa, Canada: Defence R&D Canada – Center for Operational Research and Analysis.

HMRC. (2010). Guidance on air passenger duty, what it is, who it applies to and how to pay it. HMRC website. Retrieved from http://customs.hmrc.gov.uk/channelsPortalWebApp/channelsPortalWebApp.portal?_nfpb=true&_pageLabel=pageExcise_Home. Accessed on November 24, 2010.

IATA. (2007). IATA Press Release No.: 21: Date: 04/06/2007. IATA, Vancouver. Retrieved from http://www.iata.org/pressroom/pr/Pages/2007-06-04-02.aspx. Accessed on December 9, 2007.

IATA. (2010). International Air Transport Association. Retrieved from http://www.iata.org/Pages/default.aspx. Accessed on April 16, 2010.

IATA. (2012). Climate. Retrieved from http://www.iata.org/policy/environment/climate/Pages/index.aspx. Accessed on March 20, 2012.

ICAO. (2010a). International Civil Aviation Organisation. Retrieved from http://www.icao.int. Accessed on January 6, 2010.

ICAOData. (2009). ICAOData – ICAO statistical data on the air transport industry. Retrieved from http://www.icaodata.com/default.aspx. Accessed on April 9, 2009.

IEA. (2009a). International Energy Agency Statistics. Retrieved from http://www.iea.org/stats/index.asp. Accessed on August 16, 2010.

IPCC. (2006). 2006 IPCC guidelines for national greenhouse gas inventories. Prepared by the National Greenhouse Gas Inventories Programme. H. S. Eggleston, L. Buendia, K. Miwa, T. Ngara, & K. Tanabe (Eds). IGES, Japan.

Johnston, A. (2006). Free allocation of allowances under the EU emissions trading system: legal issues. *Climate Policy*, *6*, 115–136.

Köhler, J. (2012, May 21–22). International Transport and Sustainable Development, WCTRS – Special Interest Group 2, Key Developments in the Port and Maritime Sector, University of Antwerp.

Köhler, J. (forthcoming). Globalisation and sustainable development: Case study on international transport and sustainable development. *Journal of Environment and Development*.

Köhler, J., Jin, Y., & Barker, T. (2008). Integrated modelling of EU transport policy: Assessing economic growth impacts from social marginal cost pricing and infrastructure investment. *Journal of Transport Economics and Policy*, *42*(1), 1–21.

Kyoto Protocol. (1997). Kyoto Protocol to the United Nations Framework Convention on Climate Change. Retrieved from http://unfccc.int/essential_background/kyoto_protocol/background/items/1351.php. Accessed on 6 December 2006.

Lutz, C., & Meyer, B. (2010). Environmental tax reform in the European Union: Impact on CO_2 emissions and the economy. *Energiewirtsch*, *34*, 1–10.

MacKenzie, D. (2007). Finding the Ratchet: The political economy of carbon trading. *Post Autistic Economics Review*, *42*, 8–17.

Mason, K. J. (2005). Observations of fundamental changes in the demand for aviation services. *Journal of Air Transport Management*, *11*, 19–25.

Mayor, K., & Tol, R. S. J. (2007). The impact of the UK aviation tax on carbon dioxide emissions and visitor numbers. *Transport Policy*, *14*(9), 507–513.

Montagnoli, A., & de Vries, F. P. (2010). Carbon trading thickness and market efficiency. *Energy Economics*, *32*(6), 1331–1336.

Morrell, P. (2009). The potential for European aviation CO_2 emissions reduction through the use of larger jet aircraft. *Journal of Air Transport Management*, *15*, 151–157.

Morris, J., Rowbotham, A., Angus, A.,Mann, M., & Poll, I. (2009). *A framework for estimating the marginal costs of environmental abatement for the aviation*. Omega Project 14 – Final Report, Cranfield University.

NAPs (2008). National Allocation Plans submitted by the EU Member States for the second trading period (2008–2012) of the EU ETS. Retrieved from http://ec.europa.eu/environment/climat/emission/2nd_phase_ep.htm. Accessed on July 7, 2008.

RCEP. (2002). *The environmental effects of civil aircraft in flight: Special Report of the Royal Commission on Environmental Pollution*. London: Royal College of Engineering.

SEC. (2006). European Commission staff working document: Accompanying document to the proposal for a directive of the European Parliament and of the Council amending Directive 2003/87/EC so as to include aviation activities in the scheme for greenhouse gas emission allowance trading within the Community – Impact assessment (COM (2006) 818 final SEC (2006) 1685): Unofficial advanced version, Brussels. Retrieved from http://ec.europa.eu/environment/climat/pdf/aviation/sec_2006_1684_en.pdf. Accessed on March 24, 2008.

SEC. (2008). Commission Staff Working Document: Accompanying document to the Proposal for a Directive of the European Parliament and of the Council amending Directive 2003/87/EC so as to improve and extend the EU greenhouse gas emission allowance trading system. Impact Assessment, COM(2008) 16 final, SEC(2008) 53. Retrieved from http://ec.europa.eu/environment/climat/emission/pdf/com_2008_16_ia_en.pdf. Accessed on June 11, 2009.

SRA. (2002, October). *Strategic research agenda – Volume 1*. Advisory Council for Aeronautics Research in Europe. Retrieved from http://www.acare4europe.org/docs/es-volume1-2/volume1.pdf. Accessed on March 24, 2007.

Stratus. (2005). *Controlling carbon dioxide emissions from the aviation sector*. Final Report for EPA. Stratus Consulting, Boulder, CL.

Wit, R. C. N., Dings, J. M. W., Mendes de Leon, P., Thwaites, L., Peeters, P., Greenwood, D., & Doganis, R. (2002). *Economic incentives to mitigate greenhouse gas emissions from air transport in Europe*. Delft, the Netherlands, CE.

CHAPTER 6

AIRPORT COMPANIES AS SILENT PARTNERS: THE COMPLEX INTERPLAY BETWEEN PUBLIC AND PRIVATE OWNERSHIP

Charlotte Halpern

ABSTRACT

Purpose – *Drawing on an original dataset of major European airport companies, this chapter demonstrates the growing role airport infrastructures and their managing authorities have come to play in shaping airport politics that is, how, by whom and where airports are built, modernized and expanded.*

Originality – *Airport infrastructures and companies have received little attention in recent attempts to characterize and explain the transformations of global aviation politics.*

Methodology/approach – *This chapter suggests focusing on airport companies as an attempt to characterize their long-term trajectories both in terms of their properties and in terms of their operating contexts.*

Findings – *The chapter shows that airport managing authorities have developed into full-blown economic actors, which enjoy greater levels of*

Sustainable Aviation Futures
Transport and Sustainability, Volume 4, 131–154

ISSN: 2044-9941/doi:10.1108/S2044-9941(2013)0000004006

autonomy through the systematic accumulation of resources, the diversification of revenues, and new alliances with the global finance and consulting industry. The chapter also discusses the role of privatization as the main driver for change in major European airport markets. Finally, it demonstrates the extent to which the complex interplay between public and private ownership has shaped the rescaling of the territorial dimension of airport activities, thus explaining the limited impact of anti-airport campaigns over the long-term development of major European hubs.

Implications – *This chapter has larger practical and research implications, as it demonstrates the need to go beyond a functional and context-dependent approach to airport infrastructures and managing companies.*

Keywords: Airport; airport companies; privatization; real estate; internationalization; London airports

In a context of growing competition between national economies, airports are considered essential to ensuring future economic growth and prosperity. As 'general commutators', these infrastructures play a crucial role in supporting aviation growth by attracting highly volatile flows of passengers, goods and investments. Insofar as they allow air traffic to be made operational, their expansion directly supports the competitiveness of national flag carriers and industry through constant adjustments to new challenges and issues, such as safety, technological innovation (e.g. Airbus A380) and climate change. Given their location in the vicinity of very dense metropolitan areas and their negative environmental and social impacts, airports are also aviation's most visible and contested dimension. Following a long history of policy debates on their impacts on local communities and the environment,[1] airports have developed a large range of 'sustainable aviation' policies, ranging from apparent environmental lip service to ambitious noise abatement procedures, climate action, community improvement plans and mediation centres (Walker & Cook, 2009).

Yet airports and their managing authorities, that is the airport companies themselves, have received little attention in recent attempts to characterize and explain the transformations of global aviation politics.[2] A significant share of the existing literature focuses on a range of external and domestic factors of change in order to explain evolving relationships between airlines, regulators and State authorities, such as Europeanization (Kassim & Stevens, 2010), internationalization (Thatcher, 2007) and regulatory reforms (Jordan & Schout, 2006). To be sure, airports have often been considered as mere tools in the service of the development of air traffic under the state's administrative and financial custody or at the airline's service (Rössger & Hünnermann, 1968, p. 3).[3] The development of major airport infrastructures long relied on the capacity of States and public authorities to effectively structure collective action settings. Public decisions pertaining to these instruments of national prestige were – and to a large extent still are – strongly influenced by a competitive vision of the world, which participates in the elaboration of national states interests (Hayward, 1995). However, the States' political capacity to transform their policy preferences into authoritarian actions has constantly decreased (see Kassim & Menon, 1996), as a result of the constant waning of their political, financial and technical resources since the end of the Second World War and from the unbundling of long-standing institutional arrangements between governments and industry (Héritier, Knill, & Mingers, 1996; Kassim & Stevens, 2010). This is also a consequence of the pluralization of actors and interests in a multi-level governance system (Balme & Chabanet, 2008), all of them – airlines, business groups, local authorities, NGOs – actively mobilizing in order to participate in project designs. The extent to which the loss of State autonomy is noticeable becomes even more visible when decisions regarding airport capacity and expansion are taken. In the case of international hubs such as London Heathrow (LHR), Frankfurt (FRA) and Paris Charles de Gaulle (CDG), numerous arguments have been brought forward (Halpern, 2006): local authorities denounce poor accessibility to the airport site in terms of public transport and employment, regional authorities and business groups debate the concentration of air traffic around major hubs, and last but not least, environmental groups and local communities articulate their opposition to a large range of negative externalities, such as a deterioration of property prices, the destruction of agricultural land and green belts, and air and noise pollution.

While these evolutions suggest the weakening of the State's ability to effectively regulate aviation politics by allocating resources, resolving conflicts and applying sanctions (Lange & Regini, 1989), the continued expansion of airports and air traffic across Europe suggests that other logics are at work. Building on the research that has been carried out on the limits of government (Mayntz, 1993), the chapter argues that airport expansion that is, where, how and by whom airports are built, modernized and expanded, increasingly depends on these companies' ability to effectively combine modes of state and non-state regulation. Among economists and management analysts, recent debates on airports have focused on economic regulation (Czerny, 2006; Yang & Zhang, 2011) and business strategies (Graham, 2008), and the extent to which they shape airport companies' preferences and strategies. Yet, there is a need to go beyond this functionalist approach to airports and airport companies in order to question the long-term unintended consequences these regulatory and business reforms entail for aviation politics, both in their territorial and sectoral dimensions (Margetts & Hood, 2010). Drawing on qualitative analysis and a longitudinal dataset,[4] this chapter examines how European airport companies have developed into full-blown economic actors. The comparative analysis of the long-term trajectories of three major European airports and that of their managing authorities reveals the extent to which the systematic accumulation of resources contributes to their increasing autonomy. The role of regulatory reforms as the main driver for change in major European airport markets is discussed and the extent to which the complex interplay between public and private ownership shapes the rescaling of airport activities from the European to the intra-metropolitan level of government is examined. The analysis demonstrates the need to go beyond a functional and context-dependent approach to airports and their managing authorities in order to fully understand their role in shaping sustainable aviation futures.

AIRPORT COMPANIES AS AUTONOMOUS ACTORS?

The analysis is based on the three main firms dominating civil airport management in Europe, namely Aéroports de Paris (ADP), Heathrow Airport Holdings (HAH)[5] and Fraport. Apart from managing the major international hubs of Paris CDG, London Heathrow and Frankfurt airports, these companies share a number of similarities (Tables 1a and 1b). Some relate to their origins. Originally, their main task as public enterprises was to ensure the provision (management, expansion) of sufficient airport capacity in the context of post-war reconstruction (Feldman, 1985). In doing so, they enjoyed little room for manoeuvre or innovation and were generally suffered from systematic underinvestment and low profits. Efforts first concentrated on refurbishing the infrastructures that were inherited from the pre-war period. But they also benefited from extraordinary resources, among which a monopoly over all airport infrastructures across the metropolitan area, permanent staff outposting from central government, major public investments and extensive land reserves.

Table 1a. The World's Top Passenger Hubs (Million Passengers).

2000			2012 (Estimated)		
Rank	City (airport)	Total	Rank	City (airport)	Total
1	Atlanta	80.1	1	Atlanta	95.5
2	Chicago O'Hare	72.1	2	Beijing	81.9
3	Los Angeles	66.4	3	*Lond. Heathrow*	70
4	*Lond. Heathrow*	64.6	4	Tokyo Haneda	66.8
5	Dallas/Fort Worth	60.7	5	Chicago O'hare	66.6
6	Tokyo Haneda	56.4	6	Los Angeles	63.7
7	*Frankfurt*	49.4	7	*Paris CDG*	61.1
8	*Paris CDG*	48.2	8	Dallas/Fort Worth	58.6
9	San Francisco	41	9	Jakarta	57.8
10	Amsterdam	39.6	10	Dubai Int.	58.7

Table 1b. The World's Top Cargo Hubs (Million Tonnes).

2000			2012 (Estimated)		
Rank	City (airport)	Total	Rank	City (airport)	Total
1	Memphis Int.	2.49	1	Hong Kong Int.	4.12
2	Hong Kong Int.	2.27	2	Memphis Int.	4.05
3	Los Angeles	2.04	3	Shangaï Pudong	2.97
4	Tokyo Narita	1.93	4	Anchorage	2.47
5	Séoul	1.89	5	Incheon (Corée)	2.46
6	New York JFK	1.81	6	Dubaï Int.	2.29
7	Anchorage	1.8	7	Louisville	2.18
8	*Frankfurt*	*1.71*	8	*Frankfurt*	*2.06*
9	Singapore	1.7	9	Tokyo Narita	2.01
10	Miami	1.6	10	*Paris CDG*	*1.94*

Source: Based on data provided by Airport Council International (2012).

Other similarities relate to their present situation as major economic actors. All three companies have now been privatized (although public authorities retain a majority stake in Fraport and ADP[6]). In the three cases, airport capacity and air traffic have continuously increased over time, sometimes – but not exclusively – as the result of land acquisition and infrastructure development. Notwithstanding major changes taking place in the worldwide airport network such as the rapid development of airports in South-East Asia and the Arabian Gulf, they have maintained their position among the world's ten busiest international passenger and/or cargo hubs

Table 2. Airports Companies' Main Facts and Figures (as of End 2012).

	Aéroports de Paris (ADP)	Heathrow Airport Holdings Limited (Formerly BAA Limited)	Fraport AG
Date of creation, legal status	1945, public establishment (*Etablissement public à caractère administratif*)	1985, Public limited company (British Airports Authority)	1947, Public limited company (*Verkehrsaktiengesellschaft*)
Market opening	2005	1986	2001
Main shareholders (as of end 2012)	State (54.4%), other public (29.5%), Schiphol Group AG (8%)	FGP Topco Limited, with Ferrovial S.A. (33.65%), Qatar Holding (20%), Caisse de dépôt et placement du Québec (13.29%), Government of Singapore Investment Corporation (11.88%), Alinda Capital Partners (11.18%) and China Investment Corporation (10.00%)	State of Hesse (31.4%), Stadtwerke Frankfurt a/Main (20.05%), Deutsche Lufthansa (9.98%)
Airports domestic	Paris Charles de Gaulle, Orly and Le Bourget	London Heathrow and Stansted (sold in January 2013), Glasgow, Aberdeen, Southampton	Hannover
Airports abroad (interests of at least 50%)	(Interests <50% in 12 airports)	(Last sold in 2010)	Antalya (Turkey), Lima (Peru), Burgas and Varnas (Bulgaria)
Passengers (in millions)	180 (61.1at Paris CDG)	99.7 (70 at Heathrow)	188.2 (57.5 at Frankfurt)
Total revenue	2.502 m €	2.603 m £ (thereof 241 at Stansted)	2.549 m €
Net profit	341 m €	261 m £ (t42 at Stansted)	251 m €
Employees	9,035 (thereof 8,792 in France)	8,714	20,900 (18,339 in Germany)
Net retail income per passenger	(Unknown for 2012)	5.61 £	3. 32 €
Last capacity expansion project	1989–2001 (2 additional runways in Paris CDG, terminals and intra-terminal shuttle)	1988–2008 (London Heathrow Terminal 5)	2002–2014 (Additional runway and terminal).

Source: based on data provided by ADP (2013), Fraport AG (2013) and Heathrow Airport Holdings (2013).

(see Table 2). These companies also generate an impressive annual income, a growing proportion of which is disconnected from aeronautical activities and is derived from commercial services, real estate development and international asset portfolios (see Figs. 1 and 2). While the gross revenue distribution between the companies' aeronautical versus non-aeronautical revenues shows a striking resemblance, a more detailed analysis reveals their growing specialization into different segments of airport activities with

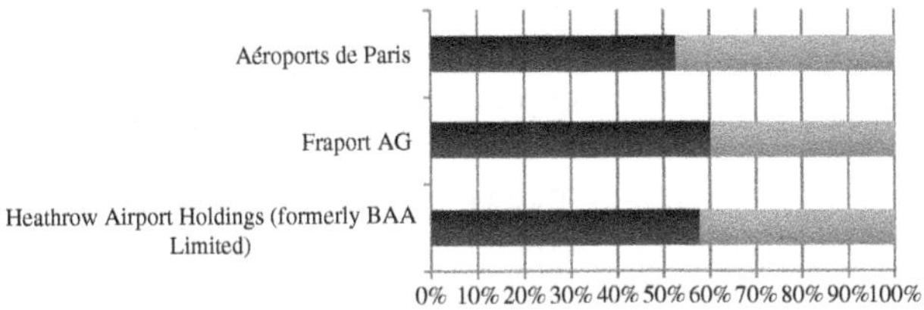

Fig. 1. Analysis of Total Revenue in € Million (as of End 2012) – Aeronautical vs. Non-Aeronautical.

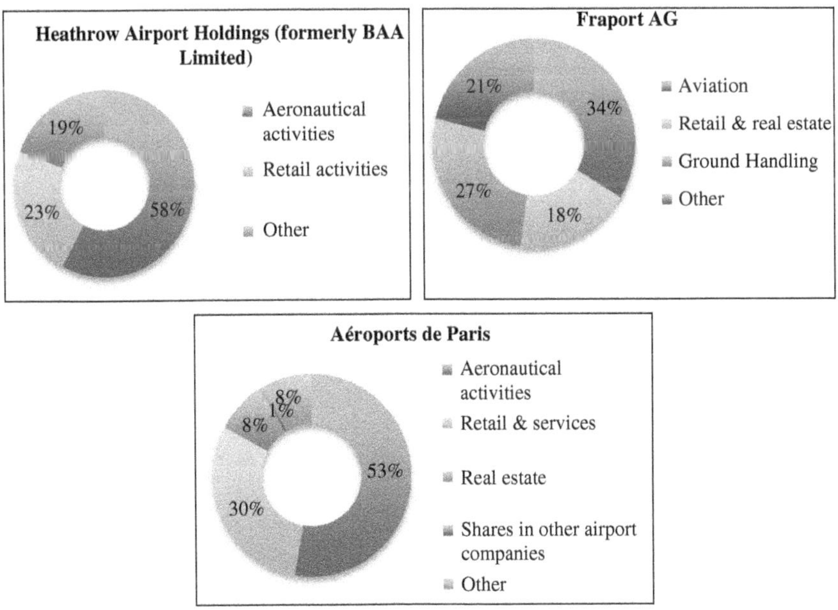

Fig. 2. Analysis of Revenue (as of End 2012) per Airport Company (in € Million).

BAA focusing on retail, ADP on real estate development and Fraport on logistics. In this sense, there has been a growing convergence towards non-aeronautical revenue streams between airport companies over time, but the degrees, rhythm and forms vary considerably between operators (Holzinger & Knill, 2005). Drawing on a comparative analysis of these companies' long-term trajectories, the chapter examines the close relationship between these dynamics of convergence and differentiation and their impact on aviation policies.

Methodologically speaking, airport companies are characterized in terms of both their properties and operating contexts. For each company, the following criteria may be considered: the major steps in the creation and evolution of the company, the origin of capital wealth, internal organization (hierarchical and territorial) and revenue distribution (by activities and based on geographical location). As for the companies' operating context, characteristics may include the national context defined by a series of instruments (rules, incentives, economic or financial instruments) as well as a more local context structured by specific networks of actors and territorial dynamics. As such, airport companies face an ensemble of constraints and opportunities within which they must operate. The properties are not significant in and of themselves, but may be used in order to demonstrate the changing nature of airport management over time, and the increasing role these economic actors play in a multi-level context.

THE CREATION AND ACCUMULATION OF RESOURCES

This section shows in more detail how the overall pattern of convergence operates in combination with logics of differentiation as observed in the management of land reserves, the development of commercial activities and strategies of internationalization. Following the 1970s economic recession and similarly to the situation observed in other parts of the world (Graham, 2008), public authorities experienced a constant decrease of their financial resources and encouraged public enterprises to diversify their income revenues regardless of their legal status. In all three cases, the first step towards greater autonomy relied on the ability to increase self-financing capacity by diversifying income revenues. All three companies engaged in this process prior to their formal privatization,[7] and in the case of Fraport and ADP, long before any privatization scheme came onto the agenda.

Land Reserves: Unequal Resources and Differentiated Strategies

Insofar as the diversification of revenue streams is closely related to air traffic growth (Morrison, 2009), the ability to draw on extensive land reserves proved to be an essential resource of the supported capacity expansion schemes. Yet land reserves are an unequally distributed resource. Some companies, such as Fraport, inherited infrastructures that were already located in densely populated areas and were soon confronted with major anti-airport campaigns (Rucht, 1984). In addition – and in spite of the initial airport size (2,100 hectares) – capacity expansion was highly constrained as the US Army occupied some 400 hectares within the airport's perimeter and located a large number of air base activities around the airport until 2005. Yet FRA's development over time shows that a poor initial endowment in terms of land reserves might lead towards alternative strategies in support of traffic growth. The company developed close links with other major German public enterprises (Lufthansa, Deutsche Post, Bundesbahn) in order to increase the airport's centrality through infrastructure networks (including high-speed rail, road and public transport). Notwithstanding the constitutional principle of decentralized infrastructure networks, these public enterprises shared an interest in the development of a major interchange infrastructure that would support the national export-oriented economic model (or so-called Deutschland AG). They consistently invested in this major gateway with the support of the German federal State, thus explaining the progressive concentration of passenger and cargo traffic at Frankfurt airport (see Halpern, 2011b).[8] This long-standing alliance still constitutes a major driver for air traffic growth and capacity expansion at Frankfurt airport: Lufthansa replaced the German federal State as one of Fraport's shareholders in 2005 and Deutsche Post co-developed an airmail terminal with Fraport in 1997. Fraport also draws on these alliances in order to increase its catchment area for both passengers and cargo traffic far beyond 200 km from the airport to include part of Belgium and the Netherlands. Other measures used to increase traffic without major capacity expansion involve air-rail services, as well as direct investments in domestic airports (e.g. Hanover). Over time, Fraport was thus able to confirm its specialization as a logistical centre and to support air traffic growth at Frankfurt airport until the departure of the US Army allowed major capacity expansion within and outside the perimeter.

Other airport companies, such as ADP and BAA, were able to develop new infrastructure further away from city centres as part of national aviation policy objectives. This strategy resulted in the construction of Paris CDG in the 1970s and London Stansted in the 1990s. In the case of BAA, London-Heathrow airport remains the United Kingdom's main hub although the original airport site (1,200 ha) cannot be easily expanded due to its location in a densely populated area. While additional capacity was provided at Stansted airport, BAA's land reserve management strategy exacerbated this initial constraint at Heathrow (and Gatwick) by systematically favouring the development of commercial space over infrastructure development. This choice led to long-term unintended effects over airport capacity management in the southeast, as discussed below. In the case of ADP and with the support of the French Department for Air Transport, the company was able to acquire an extremely large portion of agriculture land (3,100 ha). Since the new airport had not been included in the 1968 regional planning strategy, ADP enjoyed a large room of manoeuvre in designing the infrastructure and went largely unnoticed in developing a large number of commercial services and retail space within the airport site. However, this also explains why the airport's territorial integration in terms of public transport access, employment and urban planning was delayed long after its opening in 1974. Until the late 1980s, most policies, programmes and private investments concentrated around Paris-Orly airport and the newly created *villes nouvelles* (Larroque, Margairaz, & Zembri, 2002). In the absence of any form of cooperation between the 97 local authorities that were (positively or negatively) impacted by the new airport in terms of aircraft noise, job creation and income, no coherent environmental strategy was introduced until the late 1990s. In addition, ADP engaged in systematic one-to-one negotiations with each local authority in order to secure a 7,000 hectare buffer zone around the airport site through land acquisition or joint property development. Following the introduction of strict anti-noise regulation at Orly airport in 1989, ADP was able to present Paris CDG as the future 24 hour national hub against the advice of Air France, and engaged in negotiations with national and regional authorities in order to formally include the airport in the 1994 Regional Strategic Plan.

The Development of Commercial Services

Over time, capacity expansion in all three airports increasingly led to the development of commercial services.[9] BAA is generally considered to be a pioneer in this field (Graham, 2008), and its expertise was sought by a large number of airport operators throughout the world who wished to reinforce the appeal of their hub via the development of commercial activities. This strategy, which culminated in the construction of Terminal 5 at Heathrow, was initiated by Jeremy Marshall (BAA's then chief executive) and continued by his successors.[10] Both ADP and BAA developed a strategic use of the airports' modular structure in order to explore different formats and appeal to a large variety of retailers. Both started with major investments in existing terminals at Heathrow and Paris-CDG airports in order to develop commercial activities. Due to their high short-term profitability, these investments were then used to finance the refurbishment of existing installations at Paris-Orly and London-Heathrow and capacity expansion at London-Stansted, Heathrow (Terminal 5) and Paris CDG (Terminals 2E and 2F). As shown in Fig. 1, more than 40% of HAH's income now derives from non-aeronautical activities, among which a million square metres of retail space and a large variety of commercial services (including advertising, IT and parking). Similarly, Paris CDG airport is often portrayed as '*the biggest commercial centre in the northern part of the Ile-de-France region*'[11] and ADP relies on its vast land reserves both within and outside the airport's perimeter to develop a large range of commercial activities. By contrast, Fraport was unable to develop commercial services on the same level given its capacity constraint. Since 2005, 200,000 m^2 of commercial space has been developed in existing terminals and the future Terminal 3 will include 15,000 m^2 of retail space in order to accommodate the '*homo aeroportis globalis*' (Fraport AG, 2005, p. 31). Within a decade, and despite some differences in the scale of their investments, all three airports have become major 24-hour commercial centres.

The Multiple Uses of Internationalization by European Airport Companies

Internationalization constitutes a third way to accumulate resources, and this process is tightly linked to the emergence of an international airport market at the end of the 1980s. All three companies contributed to its emergence by answering capital and expertise needs to finance capacity expansion in northern America, Eastern Europe, and, later, emerging markets. As a result, they systematically developed their international asset portfolio through various types of contracts that sought to build, own, operate and/or transfer airport management knowledge elsewhere.

Different manifestations of internationalization can be observed among the main European airport companies. First, it constitutes an opportunity to create links with other large firms involved in infrastructure projects such as banks, engineering and construction groups and, more recently, the finance and consultancy industries (Lorrain, 2011). For example, Fraport invested new market opportunities in partnership with other German firms (including Dresdner and Deutsche Banks, Hochtief and Lufthansa) in Portugal, Greece and later in Russia (Moscow) and the Philippines (Manila). Second, airport companies could promote the experience they had acquired in managing their own infrastructures and, just as importantly, they could acquire and develop new competencies in other segments of the airport activity market. There again, Fraport offers an interesting example considering the constraints it faced at home. It systematically drew – and still does – on its *savoir-faire* in order to commercialize a large variety of airport services worldwide and develop new ones, particularly in the field of telecommunications and information technologies. Third, internationalization of assets also reduces the companies' dependency on domestic air traffic growth. This strategy proved particularly useful in the aftermath of the 2008 financial crisis when more than 40% of Fraport's income derived from air traffic growth and revenues elsewhere (including Turkey and Peru).

Insofar as their activities became increasingly internationalized, airport companies were also submitted to the dynamics that characterize a rapidly institutionalizing international airport market (Thatcher, 2007, p. 3). The intensification of non-aeronautical activities and the search for reliable partners abroad justified the companies' restructuring and the introduction of value management strategies. In all three cases, managers progressively replaced engineers at senior level management posts (Halpern, 2006). This process was both shaped and accelerated by processes of privatization and liberalization. This will be examined in more detail in the following section.

REGULATORY REFORMS AND THEIR IMPACT ON AIRPORT COMPANIES

Regardless of their current legal status and to the width and depth of domestic regulatory reforms, airport companies' ability to create and accumulate resources has been increasingly shaped by international dynamics. All three companies now operate in a highly competitive market in which state–market relationships have been thoroughly redefined. In addition, the growing role of global financial actors has profoundly transformed the nature of airport activities (Lorrain, 2011). This process has been exacerbated by the 2008 financial crisis. Yet their legal status and modes of action remain hybrid, as they often combine the public and private spheres in original ways as demonstrated by the long-term structuring role of historic rules and norms such as grand fathering rights on the allocation of airport slots and CO_2 emission rights.

Airport Companies: 'les maîtres de lieux'?[12]

As international hub managers, all three companies long enjoyed (and to some extent still do) a specific status at domestic and European level. As long as they operated as pioneers in an emerging market (Fligstein, 2002) and held a de facto monopoly over airport activities in their respective national markets, they were able to strategically use both internal and external resources in order shape the rhythm, width and depth of domestic and European markets. Indeed, in their extensive work on air transport in the EU, Kassim and Stevens (2010) show that the enforcement of competition rules and norms remains restricted to specific segments of airport management (e.g. ground handling services and airports slots) and within the constraints of historic rules and norms, such as grandfathering rights. The comparative analysis of three airport companies' long-term trajectories shows the extent to which they consistently mobilized in order to support – and shape – regulatory reforms at various levels of government, sometimes against the wishes of domestic regulators, policy-makers and airlines.[13] They supported the liberalization of airport services but discreetly – in comparison with airlines and air traffic controllers – resisted the enforcement of competition law by selecting pre-existing instruments (e.g. taxes, levies, grandfathering): as such they acted both as pioneers and gatekeepers in the dynamics at work in domestic airport policy change (Le Galès, 2001). Over time, all three companies reduced their dependence to national flag carriers and partly re-organized traffic flows at their main hub in order to accommodate highly differentiated types of users, including low cost airlines (Forsyth, 2007).

Recent changes at international and domestic level, as well as complaints from airport users, led domestic policy-makers to revise the first generation of regulatory reforms. Since the early 2000s, all three companies have been facing new entrants, who challenge their historic role and desire the strict implementation of competition mechanisms (Halpern & Lorrain, 2010). Similarly to the situation observed with other infrastructures and utilities

(Gomez-Ibãnez, 2003), large engineering companies such as Ferrovial, Hochtief and Vinci have invested in airport activities in search of new markets. These investments only represent a small portion of these companies' asset portfolios whether in airports or other utilities and infrastructures. More recently, financial actors such as banks and private equity funds (e.g. Macquarie Group, Qatar Holding and GIP Investments) and the larger finance industry (e.g. UBS, Goldman Sachs and Citigroup) have developed an interest in airports given both their short-term and long-term profitability. Recent changes in the British airport market offer a good example of current dynamics of change in the regulation of airport activities.

From BAA to HAH: What Implications for the British Airport Market?

Between 1987 and 2003, the development of BAA was characterized by three main features: a regime of price-based economic regulation,[14] strict monitoring of airport activities' profitability and a monopoly over London airports. This institutional setting encouraged, on the one hand, a recourse to debt financing and high returns for shareholders and, on the other, prioritized investments in commercial services. While this business strategy produced generous short-term revenues, it proved less satisfactory in a longer-term. After 2003, BAA's situation changed dramatically for both internal and external factors.

At first, BAA's business strategy proved highly effective. It provided the newly privatized company with sufficient resources to satisfy its international ambitions and to secure advantageous debt financing deals on the market. It was thus able to finance – among other things – the construction of Terminal 5 (2008), a dedicated airport rail link (Heathrow-Express, 1998) and the development of London-Stansted airport (1990s). Yet the Civil Aviation Authority (CAA)[15] repeatedly questioned BAA's apparent lack of enthusiasm in actively pursuing the construction of an additional runway at Heathrow, as part of the 1985 *White Paper on Aviation*'s and the 1996 *Strategic Economic Development Plan for the Greater London*'s implementation. Consultations on economic regulation periods show that airport users including airlines, bus companies and consumers, regularly denounced BAA's decision to strengthen its financial situation by increasing its commercial revenues instead of committing its resources to new infrastructure. These criticisms culminated with the opening of Terminal 5, where advertising and retail activities appeared to have taken over basic – yet less visible and profitable – aviation services such as luggage delivery and passenger information.

Yet over time, the strategy's long-term effects also proved highly constraining, and demonstrated both the extent and the limits of political vis-à-vis market regulation forms in shaping airport activities. After the British State renounced its golden share in 2003 under pressure from the European Commission, BAA became more vulnerable to international markets. In spite of large revenues and continued air traffic growth, the company was highly indebted and vulnerable to market fluctuations. After a long battle in which prominent actors from the finance industry such as Goldman Sachs, NM Rothschild and Citi Group played a leading role, the Spanish engineering company Ferrovial bought BAA in 2006 for £10.3 billion (£16.4 including debts). In addition, the company was less able to mobilize political and economic support in favour of infrastructure developments at Heathrow and Stansted airports. Although the 2003 White Paper on Aviation reiterated the need to expand airport capacity in the Southeast – provided it met environmental recommendations – the construction of additional runways required substantial land acquisition and the demolition of homes (Marshall, 2012).

In addition to the opposition of neighbouring communities, environmental groups and local authorities, BAA faced major criticism from the aviation industry itself regarding its interpretation of the 1986 airport economic regulation regime, and more particularly its running of Heathrow airport. While most observers recognized that the airport ran at 99% capacity, they questioned the level of taxes and fees, the allocation of airport slots and insufficient preparation for bad weather conditions. In this context, an increasing number of actors from both political and financial elites as well as the aviation industry itself called for a revision of the airport economic regulation framework (2012 *Civil Aviation Act*), which would also constitute an unprecedented opportunity to 'normalize' BAA's position and enforce competition law. Following the recommendation of the Office of Fair Trading, the Competition Commission (CC) started a market investigation of BAA soon after its acquisition by Ferrovial and recommended its dismantling in 2008. Despite the legal battle between BAA and the CC between 2009 and 2012, London-Gatwick (2009, £1.5 billion), Edinburgh airport (2012, £807 million) and London-Stansted (2013, £1.5 billion) airports were successively put on the market and sold – respectively to Global Infrastructure Partners (GIP)[16] and Manchester Airports Group (MAG).[17]

Although the company still generates considerable revenues, it faces a high level of short-term and long-term constraints due to financial, institutional and political uncertainty. In this context, Ferrovial's decision to sell 10.6% of its shares to Qatar Holding in August 2012, as well as BAA's restructuring three month later, represents the outcome of a long-term restructuring process that was initiated in 2003. The current situation of the newly created Heathrow Airport Holdings also shows the limits of a strategy that had been developed in order to comply with domestic regulatory reforms but in an entirely different political and economic context.

Who Benefits from the Dismantling of BAA's Monopoly?

Insofar as it accelerated the domestic airport market's internationalization and financialization up to a level that remains incomparable to the situation observed in continental Europe, BAA's dismantling led to a series of unexpected outcomes. On the one hand, it reinforced the position held by major international players such as Ferrovial, TBI-Abertis[18] and GIP (Table 3). But, on the other hand, it also confirms that the changing nature of airport activities is not necessarily linked to the owner's legal status. Similar to the situation observed in continental Europe, where hybrid structures such as ADP and Fraport have emerged, regulatory reforms in the

Table 3. Ownership of Major UK Airports (as of May 2013).

Ownership	Date	Current Owner (Date of Acquisition)	Airport (Rank)
100% private owned	1987	BAA-Ferrovial (40%), Qatar Holding (20%)	London-Heathrow (1), Glasgow (8), Southampton (18)
	1987	GIP (2009)	London-Gatwick (2), Edinburgh (6)
	1990	Peel Airports Ltd (1997)	Liverpool (10)
	1994	TBI-Abertis (2005)	Belfast International (13)
	2006	GIP (75%) and Highstar capital (25%) (2008)	London City Airport (16)
	1997	Metropolitan Council-Macquarie	Birmingham (7)
Mixed	2001	Local authorities + AMP	Newcastle (11)
	1987	Manchester Airports Group (thereof since January 2013: follow Manchester City 35.5%, 9 councils 29%, IMF 35.5%)	Manchester (4), London-Stansted (3), East Midlands (12),
100% public (local)	1998	TBI-Abertis (concession 30 years)	London-Luton (5)

United Kingdom also opened new opportunities for semi-private and UK-owned airport companies, such as MAG. Before the Stansted bid, Greater Manchester's ten councils owned the airport company but such an acquisition as well as future capacity expansion in both airports led to a 35.5% stake being sold to the Australian based Industry Funds Management (IMF) in January 2013. MAG has recently developed an aggressive business model in order to secure future investments and allies. In doing so, it draws on the so-called *mancunian* growth machine (Harding, Harloe, & Rees, 2010) that is, a large coalition within the Greater Manchester metropolitan area that supports a voluntarist approach to the economic development of the region and primarily relies on big infrastructure projects (inlcuding Northern Hub, Metrolink, Manchester airport and the proposed High Speed 2 railway line). Since the late 2000s, an increasing number of utilities and infrastructures were financed through equity or public–private partnerships, thus confirming a new mutation of the *mancunian* model.

BAA's dismantling also shapes domestic aviation futures. Airport capacity expansion has been on the government's agenda since the end of the 1990s, with the support of business interests and the industry. Yet the consensus regarding where, how and by whom it should be built has weakened. Until recently, the aviation industry supported a clear political decision in favour a third runway at London-Heathrow, in exchange for a clarification of competition rules in the aviation domain and a commitment by all London airports' managers to invest in airport infrastructure during the next economic regulation period. In a context of austerity, other actors are pushing for alternative solutions such as high-speed rail projects and effective airport decentralization in support of regional economic growth (see, e.g. the 2012 *Heseltine report*).[19] By highlighting the negative long-term effects of air traffic concentration in the Southeast, representatives from the 'northern region' have suggested a clear relationship between sustainable aviation futures and the strengthening of city-regions outside London.[20]

On the whole, recent changes in the UK airport market help reconsider the role of privatization as the main driver for change. It also reveals the complex interplay between public and private ownership and between state and market modes of regulation. Indeed, airport companies are growing increasingly international thanks to their improved financial status. They intervene simultaneously in both market and non-market spheres by drawing on both regulatory and economic tools. As a result, their long-term strategies are jointly shaped by internationalization and financialization dynamics on the one hand and differentiated political capacity on the other hand. Similarly to the situation observed in France, Germany and beyond, discussions regarding the economic impact and contribution to growth of a country's main hub are increasingly addressed as part of a wider reconfiguring of central–local relations. This evolution confirms the importance of a third dynamic at play that is, territorial dynamics, which will be addressed in the next section.

THE RESCALING OF AIRPORT MANAGEMENT

In a context of increased financial and economic uncertainty, airport companies have paid renewed attention to their territorial integration in order to reactivate or create new partnerships with public authorities at all levels of government. Indeed, their active support through attractive strategic planning, economic development, transport and housing policies, programmes and investments, has proved essential in order to develop a third generation of airport infrastructures that is, airport city projects. This '*combination of giant airport, planned city, shipping facility and business hub*' (Kasarda & Lindsay, 2011, p. 2) intensifies revenue maximization strategies up to a level that remains incomparable to the situation observed at 'regular' hub facilities. To be sure, the development of airport city projects contributes to further de-territorializing airport activities, as they are made less dependent on their neighbouring territories in terms of workforce, passengers and goods. Unlike hub infrastructures that had to be located near large metropolitan areas, airport city projects rely on their accessibility, land reserves and cash flow. In the case of the three airport companies under scrutiny in this chapter, this evolution has larger implications in terms of territorial rescaling (Brenner, 2004). More precisely, it can be seen to exacerbate the re-concentration of European, national and subnational public and private investments, programmes and policies around a small number of strategic infrastructures.

In a context of air traffic re-concentration, historic alliances between airport authorities and national flag carriers have been re-enacted in order to keep and increase market shares within the limits set by competition law. Yet these alliances now occur at the EU level and in a far more integrated form than international airline alliances. Lufthansa offers a good example of this recent evolution. It relies primarily on other passenger airlines (including Swiss, Brussels Airlines, Austrian Airlines and Germanwings) it has acquired through mergers, takeovers and acquisitions but as Fraport's shareholder, the Lufthansa Airline Group also promotes a similar integration of Frankfurt, Munich and Zurich airports. Following the merger between Air France and KLM in 2004, ADP and Amsterdam's Schiphol Group exchanged 8% of their respective shares in order to create a major hub alliance. The functioning of hub alliances also depends on increasing the density of European high-speed train networks for both passengers and cargo as supported by Trans-European Transport Networks programmes, which provides direct connections with other major infrastructures across Europe such as ports (i.e. Rotterdam, Hamburg and Le Havre) and economic development zones far beyond the airports' current catchment areas.

Airport city projects also have profound implications at (intra-)metropolitan levels. At this scale of intervention, airports' territorial integration was long associated with sustainability policies and conflict management. These initiatives, ranging from symbolic towards more substantial measures, aimed to involve neighbouring communities in an airport's day-to-day management through employment, economic development and anti-noise procedures. Yet airport city projects extend their managers' influence far beyond the limits of airport sites as they become directly involved, as investors or as developers, in large-scale real estate, transport and economic development at the metropolitan level. These developments have major implications on sustainability and aviation's carbon footprint at a metropolitan scale in that they rely on an extensive and systematic exploitation of land, energy and natural resources around airport sites, and endanger these metropolises' green infrastructures (e.g. greenbelts and ecological corridors). To be sure, airport companies were already involved in such projects outside the airport perimeter, as shown by their active role in developing parking, hotels, convention centres and office space. In the case of BAA and ADP, this had been the starting point for investments in central Paris and London as well, with BAA's involvement in Marylebone and St Pancras an example of this approach. Yet airport city projects tend to intensify this logic up to a new level, in terms of project scale and penetration in urban areas. ADP offers a good example of this long-term strategy to specialize into real estate and property development. Over time, it has accumulated thousands of square meters of land and property throughout the Parisian metropolitan area, and simultaneously launched two airport city projects (Coeur d'Orly and Triangle de Gonesse) in partnership with public authorities and real estate development firms.

Subnational authorities are often directly involved in these schemes as airport companies' owners, project investors or project partners. This is the case of Frankfurt City, which is involved in all three capacities with Fraport and local real estate firms in the development of Gateway gardens, a new business district located on the fringes of the airport. The Manchester 'Airport City' plan follows a similar strategy as part of the creation of a business district in the Wythenshawe district, an area that is characterized by a high level of (especially youth) unemployment. Whether or not local neighbourhoods will benefit from these developments' economic impact in terms of increased employment, investment, income and tax revenue is highly uncertain as shown by debates among economists regarding the establishment of formal causality between airport expansion and wider economic development (see Graham, 2008). However, airport city projects exert major territorial and political implications on the allocation and distribution of public policies, investments and programmes in a large variety of policy domains such as economic development, housing, public transport education and training. Airport companies have actively engaged in various economic and political fora at the regional and metropolitan levels in order to shape decisions regarding the location of major investments and infrastructures, as well as the elaboration of strategic planning documents. This is also the case of research and development policies, such funding being justified in view of carbon footprint reduction and contribution to climate mitigation. Yet by engaging in partnerships with universities, utilities companies, government agencies and local authorities, airport companies also seek to reduce energy consumption and waste production through green growth technologies.

Insofar as they shape the restructuring of state spaces in Europe by contributing to its further centralization around a small number of infrastructures, evolving relationships between airport companies and their respective territories confirm the rescaling of airport companies' strategic capacities. It shows that airport companies remain heavily dependent on public intervention and resources, whether to retain their strategic position or to access new markets, develop their activities and reduce negative externalities. This somewhat contradictory process explains their attempts to simultaneously shape social mobilization and power relations at various levels of government.

CONCLUSION

Drawing on a comparative analysis of the three main European airport companies, this chapter demonstrates the need to include airport infrastructures and their managing authorities in the study of sustainable aviation futures. While both practitioners and academics had long considered these actors as silent partners, they have developed into full-blown economic actors, which enjoy significant levels of autonomy. In Paris, London and Frankfurt, ADP, BAA (now HAH) and Fraport have engaged in a resource maximization strategy that goes far beyond airport site limits with the support of public authorities at all levels of government. The analysis presented here contributes to current debates on the limits of government and shows how sectoral, political and territorial dynamics jointly shape airport capacity expansion thus explaining variations in time and space. In other words, convergence occurs as a result of both differentiation and standardization mechanisms. This has major implications for sustainable aviation futures, inasmuch as it shapes political capacity restructurings in its ability to effectively regulate airport infrastructures.

NOTES

1. For an overview, see the general introduction to this edited volume.

2. The social movements literature has devoted considerable attention to anti-airport campaigns, but their long-term implications on airports' trajectories and aviation policies has often been neglected (Halpern, 2006).

3. A similar claim was recently made regarding control and command services, such as slot allocation in the aviation sector, necessary for monitoring access to and optimizing the uses of the infrastructure (Csikos & Pflieger, 2012).

4. Data has been gathered in the three cases according to the principle of triangulation between secondary sources, documentary sources and semi structured interviews with different representatives from airport companies, and with other private, public and social actors involved in airport planning policies – either as projects' promoters or opponents. We also did a systematic review of the business press in all three countries though Factiva and LexisNexis databases and of the companies' annual reports since the early 1990s. The analysis has been developed in more details in a series of monographs on each firm (Halpern, 2011a, 2011b; Halpern & Lorrain, 2010), as well as in a doctoral thesis on contentious airport planning in France and Germany (Halpern, 2006).

5. Heathrow Airports Holding Limited (formerly BAA limited) was created in October 2012. Since 2006, it is owned by FGP Topco Limited, an international consortium led by the Spanish engineering group Ferrovial. Following the group's dismantling, a profound reorganization of HAH's structure and debt financing took place in 2012.

6. Until the 2008 financial crisis, British airports remained an exception in the EU in terms of both privatizations' degree and financing strategies. For a worldwide review, see Graham (2008, pp. 18–36).

7. Other cases in Europe confirm the absence of a direct relationship between the development of commercial activities and privatization (see Csikos & Varone, 2013).

8. The German Airports Association provides extensive longitudinal data on domestic air traffic: http://www.adv.aero (June 20, 2013).

9. 'Retail revenue per passenger' is now considered a major indicator of a company's financial reliability (see Tables 1a and 1b).

10. Interview with former BAA senior executive, London, June 2010.

11. Interview with urban planner, Paris, March 2003.

12. Le Galès (2001, p. 137).

13. See recent debates in all three countries regarding the economic regulation of airports.

14. All three London airports and Manchester airport operate under a single-till cap economic regulation regime. According to Czerny (2006, p. 86), '*the key difference between single-till and dual-till regulation is that with single-till the price-cap is set in anticipation of the revenues from aeronautical and commercial services*'.

15. A thorough review of the financial press and of the main protagonists' annual reports during the 1990s gives a good overview of the long struggle between the CAA and BAA. This was later confirmed in interviews with representatives from both organizations. The CAA was created in 1972 as the main regulator for all aspects regarding civil aviation in the United Kingdom. This independent agency has no equivalent in the French and the German context, where, respectively, the Department for Air Transport and Regional authorities ensure this function.

16. GIP is an American private equity fund that was created in 2006. It holds stakes in a large number of utilities and infrastructures across the world.

17. MAG already owned Manchester, East Midlands and Bournemouth airports in the United Kingdom.

18. This international group is partly owned by Aena – the Spanish airport company. It holds stakes in 30 airports and 9 countries, with a global traffic of 80 million passengers a year.

19. The Cameron government, as well as the conservative party, are profoundly divided on this issue.

20. See also the recent discussion on the 2013 Draft Aviation Policy Framework.

ACKNOWLEDGEMENTS

The research for this chapter was funded by the Sciences Po Cities Program, as part of a series of monographs on European airport companies. The author is particularly thankful for the comments from the book's editors and participants to the Sustainable Aviation Workshop at the University of Essex in March 2012.

REFERENCES

ADP. (2013). Document de référence et rapport financier annuel 2012. Paris: Aéroports de Paris. Retrieved from http://www.aeroportsdeparis.fr/. Accessed on June 20, 2013.

Airport Council International (2012). Annual traffic data. Retrieved from http://www.aci.aero/Data-Centre/Annual-Traffic-Data. Accessed on June 20, 2013

Balme, R., & Chabanet, D. (2008). *European governance and democracy*. New York, NY: Rowman and Littlefield.

Brenner, N. (2004). *New State spaces*. Oxford: Oxford University Press.

Csikos, P., & Pflieger, G. (2012). Regulation of liberalised network industries. *Swiss Political Science Review*, *18*(4), 428–451.

Csikos, P., & Varone, F. (2013). Impacts de la libéralisation sur les aéroports nationaux en Suisse. *Flux*, *92*, 6–22.

Czerny, A. (2006). Price-cap regulation of airports. *Journal of Regulatory Economics*, *30*(1), 85–97.

Feldman, E. J. (1985). *Concorde and dissent*. Cambridge: Cambridge University Press.

Fligstein, N. (2002). *The architecture of markets*. Cambridge: Cambridge University Press.

Forsyth, P. (2007). The impacts of emerging aviation trends on airport infrastructure. *Journal of Air Transport Management*, *13*, 45–52.

Fraport AG. (2013). *Annual report 2012*. Frankfurt a/Main, Germany: Frankfurt Airport Services Worldwide. Retrieved from: http://www.fraport.com/. Accessed on June 20, 2013.

Gomez-Ibãnez, J. A. (2003). *Regulating infrastructure*. Cambridge, MA: Harvard Unviersity Press.

Graham, A. (2008). *Managing airports* (3rd ed.), Oxford: Elsevier.

Halpern, C. (2006) *La décision publique entre intérêt général et intérêts particuliers*. Thèse de doctorat. Paris: IEP de Paris.

Halpern, C. (2011a). La gestion aéroportuaire a-t-elle changé de nature? *Flux*, *83*, 68–82.

Halpern, C. (2011b). Bâtisseur, gestionnaire et aménageur: les stratégies adaptatives de Fraport AG et Aéroports de Paris. *Flux*, *84*, 73–89.

Halpern, C., & Lorrain, D. (2010). Portraits d'entreprise: Hubs (ports et aéroports). *Flux*, *79-80*, 140–152.

Harding, A., Harloe, M., & Rees, J. (2010). Manchester's bust regime. *International Journal of Urban and Regional Research*, *34*(4), 981–991.

Hayward, J. (Ed.). (1995). *Industrial enterprise and European integration*. Oxford: Oxford University Press.

Héritier, A., Knill, C., & Mingers, S. (1996). *Ringing the changes in Europe*. Berlin: de Gruyter.

Jordan, A., & Schout, J. A. (2006). *The co-ordination of European governance*. Oxford: Oxford University Press.

Kasarda, J., & Lindsay, G. (2011). *Aerotropolis*. London: Allen Lane.

Kassim, H., & Menon, A. (Eds.) (1996). *The European Union and national industrial policy*. London: Routledge.

Kassim, H., & Stevens, H. (2010). *Air transport and the EU*. Basingstoke, UK: Palgrave Macmillan.

Holzinger, K., & Knill, C. (2005). Causes and conditions of cross-national policy convergence. *Journal of European Public Policy*, *12*(5), 775–796.

Lange, P., & Regini, M. (1989). *State, market and social regulation*. Cambridge, UK: Cambridge University Press.

Larroque, D., Margairaz, M., & Zembri, P. (Eds.). (2002). *Paris et ses transports (XIX° et XX° siècles)*. Paris: Recherches.

Le Galès, P. (2001). Est maître des lieux celui qui les organise. In A. Stone Sweet, W. Sandholtz, and N. Fligstein (Eds.), *The institutionalization of Europe* (pp. 137–154). Cambridge: Cambridge University Press.

Lorrain, D. (2011). The discrete hand. Global finance and the city. *Revue Française de Science Politique (English)*, *61*(6), 1097–1122.

Margetts, H., Perri, G., & Hood, C. (2010). *Paradoxes of Modernization*. Oxford: Oxford University Press.

Marshall, T. (2012). *Planning major infrastructure*. Oxford: Routledge.

Mayntz, R. (1993). Governing failures and the problem of governability. In J. Kooiman (Ed.), *Modern governance* (pp. 2–20). London: Sage.

Morrison, W. G. (2009). Real estate, factory outlets and bricks: A note on non-aeronautical activities at commercial airports. *Journal of Air Transport Management*, *15*, 112–115.

Rössger, E., & Hünnermann, K.-B. (1968). *Einführung in die Luftverkehrspolitik*. Manheim, Germany: Bibliographisches Institut/Hoschultaschenbücher – Verlag.

Rucht, D. (Hg.) (1984). *Flughafenprojekte als Politikum*. Frankfurt: Campus Verlag.

Thatcher, M. (2007). *Internationalisation and economic institutions*. Oxford: Oxford University Press.

Walker, S., & Cook, M. (2009). The contested concept of sustainable aviation. *Sustainable Development*, *17*, 378–390.

Yang, A., & Zhang, A. (2011). Price-cap regulation of congested airports. *Journal of Regulatory Economics*, *39*(3), 293–312.

CHAPTER 7

SUSTAINABILITY AND NOISE ANNOYANCE

Christian Bröer

ABSTRACT

Purpose – *This chapter addresses the political history of aircraft noise annoyance and its relationship to sustainability.*

Methodology/approach – *Using extensive case studies and secondary sources, the chapter shows how a particular approach to noise annoyance that emerged already accepted large-scale annoyance. The focus is directed towards the black boxing of political decisions with scientific knowledge.*

Findings – *This contribution demonstrates that noise annoyance policy has shaped citizens' perceptions of aircraft sound and has intensified the associated annoyance. Finally, the discussion considers the significance of the participation of citizens and the political implications of these findings. Currently, participation is predominantly confined within a growth paradigm and reproduces tensions in ecological modernization policies. Participation and protest can inadvertently engage these tensions and question the participatory practices themselves. A moratorium on*

Sustainable Aviation Futures
Transport and Sustainability, Volume 4, 155–173

ISSN: 2044-9941/doi:10.1108/S2044-9941(2013)0000004007

'growth' might create the necessary space for fundamentally rethinking aviation and sustainability.

Keywords: Noise annoyance; sustainability; policy; protest; complaints; participation

Sustainable aviation is a highly contested concept (Walker & Cook, 2009). Attempts at 'balancing' aviation growth and environmental impact with an ecological modernization approach appear to offer no reconciliation (Boons, van Buuren, & Teisman, 2010; Bröer, 2006, 2007; Huys, 2011; Kroesen, 2010; van Eeten, 1999; Walker & Cook, 2009). In many countries, noise annoyance from civil aviation serves as a yardstick for the negative impacts of air transport and, as Thomas and Lever (2003, p. 97) suggest, 'It is widely accepted that the most significant local environmental impact associated with the operation of airports arises from noise generated by the aircraft'.

As this chapter will show, the seemingly apparent relationship between aviation noise and sustainability is a historically specific intervention. This chapter looks at the particular ways that sustainability is approached at two major European airports – Amsterdam Schiphol and Zurich Kloten. In Amsterdam especially, noise annoyance has become the central focus in an ecological modernization approach for airport growth. To demonstrate the specificity of this approach, it is contrasted with the case in Zurich.

A desire to minimize noise annoyance has been used as a strategy to limit the growth of air transport. However, this historically and regionally specific combination can be questioned, firstly for its effectiveness and secondly for its unintended side effects such as postponing CO_2 reductions or sensitizing local airport communities to aircraft noise. In the case of Schiphol, approaching sustainability through noise annoyance has not reduced the overall environmental impact of the airport's operation. It did, however, alert citizens to the issue of aircraft noise and increase annoyance.

Noise annoyance caused by air transport operations emerged as a major political problem after World War Two. Annoyance at aircraft noise had been coupled with sustainability since the early 1980s, but the focus has recently shifted to aviation's other environmental impacts. It is telling that Earthscan's second book on sustainable aviation no longer deals with issues of aircraft noise and community conflicts (Gössling & Upham, 2009; Upham, 2003). In the past, Friends of the Earth (FoE) used noise annoyance policy to erect barriers against industry expansion and thereby attempted to limit its growth. In a recent report, the argument is different: 'People living near airports have long suffered from aircraft noise, traffic congestion and air pollution. Indeed communities around airports have been concerned about these issues for years. However, new evidence shows that air travel is contributing towards a far greater threat – CLIMATE CHANGE' (FoE, 2012).

It is time to look at the relationship between sustainability and noise annoyance in a critical way. This chapter first addresses the political history of aircraft noise annoyance and its relationship with sustainability. Secondly, it demonstrates that noise annoyance policy has shaped citizens' perceptions of aircraft sound and intensified annoyance. The way citizens are implicated in annoyance policies will be addressed in the third section. Finally, the political implications of these findings will be discussed.

THE PUBLIC ISSUE OF AIRCRAFT NOISE

When humans are exposed to acoustic stimuli they react in diverse ways, both individually and historically. Sounds, for example of church bells, minarets, carriages or cars, can be experienced as civilization or its discontents (Bailey, 1996; Schafer, 1994). Levels of acoustic energy, expressed in decibels (dB(A)) or otherwise, rarely explain peoples' reactions (Fidell, 2003; Job et al., 1996; Maris, Stallen, Vermunt, & Steensma, 2007; Schultz, 1978; Stallen, 1999). Today, less noise is needed to annoy people (Bröer, Wirth, Brink, & Schierz, 2005; Schreckenberg & Meis, 2007; van Kempen & van Kamp, 2005).

Which sounds get onto the public agenda, and how problems of noise are framed, differs by sound source, time and place. Responses to neighbourhood noise in Europe, for example, followed a completely different trajectory than that of traffic or industrial noise in the 19th and 20th centuries (Bijsterveld, 2008). While neighbourhood noise was approached as a matter of individual sensitivities and mediation, industrial noise was approached as a matter of public health concern in Western Europe and the United States. In China, in contrast, traffic noise is related to harmonious interaction. It is clear then that noise annoyance needs to be set in the context of historically grown political agendas.

Aircraft sound emerged as a serious public problem in the Global North after World War Two (Dierikx & Bouwens, 1997; Heppenheimer, 1995). This chapter describes the detailed histories of aircraft noise annoyance at

Amsterdam's Schiphol airport in the Netherlands and Zurich Kloten airport in Switzerland, paying particular attention to how science became involved in policy.

NOISE POLICIES AND RESEARCH IN AMSTERDAM AND ZURICH

For the last seven decades or so, the development of aircraft noise policies has involved and steered scientific research. Noise policies acquired a scientific quality and legitimacy in the sense that acoustic, health and 'community noise' research was mobilized frequently in political decisions and in policy implementation. Aircraft noise contours and noise metrics are a perfect example of this. In turn, noise research has become politicized. This means, firstly, that it has become partisan: in conflicts over noise exposure, different parties mobilized different knowledges. Secondly, it refers to the fact that research has became accountable to political choices and assumptions. Research is likely to flourish and gain official legitimacy when it fits into dominant political agendas. Thirdly (and partly crosscutting the first trend), historical choices become taken for granted and institutionalized over time and were no longer substantially questioned. For almost all parties involved in noise annoyance politics, the founding assumption was the perceived socio-economic necessity of air transport mobility and growth. Therefore, it is essential to consider this assumption and then examine how it has affected the relationship between science and policy in the cases of Amsterdam and Zurich airports.

Commercial air travel is now over a hundred years old, although large-scale mass civil aviation didn't emerge until after World War Two. In Amsterdam, Schiphol airport had to be rebuilt after the war, while in Zurich a completely new airport was constructed at Kloten (Dierikx & Bouwens, 1997). In both cases, policy-makers presented their airport development plans as being necessary and in accordance with the dominant discourse of aviation expansion. In many countries, civil airports still had to be built. The necessity to construct a passenger airport was an expectation at best. In a range of policy documents from the post-war period, it is apparent that both national governments and the aviation industry presented air transport development as a natural process of economic progression. This expansionist argument was rarely challenged in the past and it remains dominant in airport policy in many countries today.

The growth trend argument was institutionalized in policy related science. From the beginning of aircraft noise policy in the 1950s, policymakers in Europe and the United States turned to scientists to estimate (future) levels of community annoyance. In many cases, scientists worked out noise annoyance criteria that incorporated the trend argument and permitted future aviation growth (Bijsterveld, 2008; Bröer, 2007). Noise annoyance limits are a clear example of this: no airport is ever large enough in area to fully prevent annoyance or health impacts. Consequently, while all noise exposure limits attempt to curtail community annoyance, they also accept that a certain degree of annoyance is almost inevitable.

By and large, policy related annoyance research is based on the following logic. First of all, noise annoyance is understood to be a subjective phenomenon and, specifically, it is considered to be a matter of personal perception. Secondly, average perceptions for groups are introduced and objectified by correlating them with sound exposure levels. These sound exposure levels become key factors for policy even though, statistically, levels of acoustic energy can only partly predict resulting levels of annoyance. These metrics become relevant because sound exposure levels appear to fit political ambitions to manage annoyance by 'turning down' the volume. It can be understood as being a modernist technocratic approach insofar as it assumes that the object of policy is a more-or-less uniform population. From this perspective, scientific research has often approached the relationship between noise perception and exposure and repeatedly produced similar results. Different metrics for noise exposure (such as average noise exposure over time, e.g. L_{eq}) and intricate models to predict exposure are the byproducts of a political choice to approach aircraft noise annoyance in this particular way.

In both the Netherlands and Switzerland, as in many other countries, aircraft noise annoyance limits are based on the average annoyance score in a chosen area. In this way, the problem of aircraft noise annoyance acquires a spatial quality. As a result, a substantial number of people beyond the annoyance limits may be highly annoyed, while many inside the contours are not. In fact, the majority of annoyed citizens often live outside of the areas are addressed by noise policy. In other words, the annoyance policy and its corresponding underpinning science are based on the assumption that annoyance is unavoidable to a large degree.

While alternative approaches that did not presuppose or complacently accept this assumption were used in the very early years of noise policy and in the margins of noise protests, they did not gain much currency. Dutch scientists who commissioned research into aircraft noise annoyance around

Schiphol in the early 1960s suggested that a different runway layout should be developed to avoid any contact with urban areas. In Switzerland in the 1950s, lawyers argued for 'silence as a basic right' as opposed to silence as a relative interest while, several years ago, residents living near London Heathrow asked the European Court for Human Rights to rule that night noise breached their right to sleep.

In policy, large-scale annoyance is implicitly accepted as part of the trend argument. Politicians, for their part, legitimize noise policy with reference to science-based criteria; however, scientists have not successfully established productive acoustic definitions of noise annoyance. An analysis of early noise annoyance research shows the same results for both Amsterdam and Zurich. Scientists were able to present correlations and distributions of noise annoyance over particular residential areas or groups of residents, but they could not determine how much noise was too much. Nevertheless, scientists determined arbitrary limits by incorporating already existing policies into their methods and reports.

Bijsterveld (2008) was able to unlock the black box of scientific expertise by studying the archives and mail exchanges between some of the chief scientists of that time – Kosten, an acoustician, and Bitter, a psychologist. Kosten was the head of the committee that was assigned the task of devising a Dutch noise index and presenting a scientifically based threshold of annoyance. In other words, he was responsible for resolving the question of how much noise was too much. Bitter, the psychologist, was held responsible for providing the answer. Bitter, however, replied that he could only show correlations and could not designate what was normatively right or wrong. It took months of pressure from the committee before Bitter agreed to formulate some thresholds that Kosten could translate into limit values to inform policy. Bijsterveld presents this as an exemplary instance of 'pragmatic objectivity', the logic of intervening with the aid of numbers. A certain acoustic load as a sign of annoyance is now predominantly taken for granted, and yet this essentially basic 'science', which has become embedded in policy, had to be forced into scientific practice.

The last similarity between the two cases concerns the timing of noise annoyance policy. In Zurich and Amsterdam, the trend argument preceded attention being given to the harmful effects of air traffic. Policy documents first spoke about airport enlargement and economic benefits (1945–1955) and only later (from 1955 onwards) included noise and other concerns. This was even the case for successive plans to enlarge the airports or alter policy: repeatedly, policy-makers were first concerned with the economic benefits of expansion and only later considered other concerns.

AMSTERDAM: ECOLOGICAL MODERNIZATION

Differences between the two cases are apparent in relation to the way noise annoyance was first introduced. In the Netherlands, airport planners, experts and policy-makers introduced the noise issue themselves, and policy-related science remained central for decades. In Switzerland, on the other hand, citizens played a more active role. In Amsterdam, the first plans for airport enlargement contained no reference to noise. The first report to mention noise annoyance dates from 1955 and was written by municipal planners. At the time, engineers, policy-makers and scientists were devising long-term plans for Amsterdam and The Netherlands and stated, 'The problem of annoyance has become more current all over the world'. This referred to recent conflicts over airport noise that had occurred in New York. As part of their planning practice, they envisaged the future possibility of conflicts between housing development and aircraft flight paths. In recognition of this, they called in scientists to estimate future noise levels and associated community annoyance (Bijsterveld, 2008; Dierikx & Bouwens, 1997). As a consequence of this approach, annoyance became a central component of long-term spatial planning practice.

When aircraft noise was first introduced as an issue, experts reverted to the existing and widely shared planning practice to tackle it. In spatial planning, annoyance was defined as a national problem, which could be solved by central planning policy and noise limits for certain areas. 'Experts' figured prominently in the planning debate and local airport residents were approached as a standardized and passive population. Planning was done for them and noise annoyance was treated as an averaged quantity. Noise production was thus accepted as an inevitable part of air traffic growth but its exposure had to be managed. The noise issue was first discussed among planning professionals at the airport before spreading as a topic of debate among politicians, the media and the public (Dierikx & Bouwens, 1997).

In 1960, about five years after the start of the noise policy at Schiphol, individual citizens started complaining and petitioning local politicians and airport officials. Citizens and protest groups increasingly advanced proposals that undermined the current policy. Most notably, they pleaded for the total relocation of Schiphol. At the same time, they further strengthened the dominant discourse: most activists subscribed to the acoustic and technocratic model of spatial planning, noise limits and noise contours. At the end of the 1970s and the beginning of the 1980s, the planning approach confronted a deadlock: noise limits or expansion. Eventually, neither the noise limits nor the plans for a new runway were implemented.

The situation changed with the advent of ecological modernization. From the 1980s onwards, the planning approach became mixed with ideas from ecological modernization (Hajer, 1995). Ecological modernization comprised a new, positive-sum logic: a promise from policy-makers to expand the airport while, crucially, alleviating noise exposure at the same time. From this perspective, one could simultaneously increase air traffic and lower the airport's environmental impact. Noise exposure and airport expansion were framed in terms of ecological modernization for the first time in 1986 in a report on economic development commissioned by the central agency for spatial planning (again!). The document called Schiphol a 'mainport' for the first time. This term, similar to the term hub, was invented and used to refer to both Schiphol and the harbour at Rotterdam. It summed up the older trend argument and became the central storyline of airport policy. Within the frame of ecological modernization, the proposed new runway was reframed as an 'environmental runway' since it was supposed to handle flights with less noise exposure. Ecological modernization came with a repertoire of similar phrases: 'sustainable airport growth', 'moderate growth' and 'conscientious growth'. 'Mainport and ecology' were thus institutionalized as the leading idea of the central government in numerous policy plans to date (Huys, 2011).

Ecological modernization made it possible to start building the new runway and implementing noise limits. Noise, in this context, was seen as an environmental problem (and less as a health problem, for example) and became the yardstick for evaluating the negative effects of airport growth. This was not directly triggered by the environmental movement. To the contrary, Friends of the Earth Netherlands used this opportunity to instigate acts against airport noise in the early 1990s, at least 5 years after the noise problem was redefined as an ecological issue in national policy. FoE deliberately chose to act on noise annoyance since they assumed that limiting noise would limit the growth of Schiphol. They successfully acted as an umbrella organization of local anti-noise groups and became the most important stakeholder organization and entered into negotiations with central government and the aviation sector.

The policy had created an entry point in three ways: firstly, it reframed sound as an environmental problem; secondly, it further stressed national policy as the arena for addressing this issue. Therefore, the national environmental movement was able to put itself at the head of several local protest groups, which had previously been more concerned with local living conditions. Thirdly, it reflected a change in policy-making. In the 1980s, planning procedures gradually became 'interactive'. In the case of Schiphol and airport enlargement, this had several implications: citizens were informed about airport policy plans extensively and early in the process; government, industry and protest groups told citizens that they had a stake in airport policy and they all pointed to noise as a source of (future) harm. Furthermore, complaint facilities and inquiry procedures were set up or enlarged and presented to the public. Citizens were invited to take part in 'interactive policy-making'. The government redefined protest groups as stakeholders and invited them for consultation.

The new discourse of ecological modernization facilitated policy-making in the sense that it solved the contradiction between airport expansion and noise mitigation. Both were redefined as part of one, ecologically modern, policy. On the other hand, the same policy further emphasized noise as the central problem of airport growth. Following the Dutch policy concepts, I have termed this the 'mainport and environment discourse', which became the dominant discourse from 1991 onwards. Central government's attempt to get more citizens involved in the policy process gave rise to a contradictory situation: noise policy was still based on expert knowledge, standard annoyance reactions, central planning and a passive population yet the process itself mobilized social movements, individual citizens and their own judgements.

SWITZERLAND: NOISE REDISTRIBUTION

In contrast with the situation in the Netherlands, citizens in Switzerland were more important from the outset, and the 'scientization' of noise policy was more limited since noise remained more directly politicized. Noise annoyance at Zurich airport was mentioned for the first time in 1965 in the national parliament, which immediately appointed an expert commission on noise annoyance (Gallati, 2002). Members of the parliament had been steered by the newly founded Swiss League against Noise. In 1955–1956, two jurists from Zurich joined in protest against several noise sources and asserted a basic right to silence in their first declaration. The League insisted on quietness as a human right. They portrayed technology as a threat to human freedom that had to be tamed by democratic legal rule. Members of the League were active members of the ruling party and high-ranking professionals. Their concerns were soon taken up in federal policy, but not to their satisfaction. The government appointed its 'Federal Expert Commission' in 1956. In the Commission, policy-makers, industry and experts held a majority. Most importantly, quietness was reframed as one among many different interests such as economic benefits, for example. The Commission set for itself the task of balancing those different interests, diverging from the idea of a 'basic human right'. The first report of the Commission in 1963 defines noise annoyance as opposed to other interests in that it is an 'inevitable conflict'. This conflict, the Commission stated, should not be solved as such but ought to lead to a compromise (Eidgenössisches Expertencommission, 1963, pp. 144–145).

The expert commission was appointed by the Federal Government. Later, the Regional Government (Canton) followed with a commission of its own. Municipalities, and even villages, developed their own individual noise policies. The different layers of policy both aligned and diverged on the noise issue and this complex interaction is an ongoing feature of noise policy at Zurich Kloten airport. Annoyance was 'federalized' and approached as an issue of federal policy-making from the outset. This means that national, regional and local governments are responsible at the same time; responsibilities are negotiated repeatedly; and citizens participate in the policy process through referenda and by directly pressuring local politicians. As in the Netherlands, where noise was approached as a planning issue, policy-makers in Switzerland in the 1950s employed an already existing discursive practice to address the upcoming issue of aircraft noise. This 'federalization' of noise meant that local communities were strongly involved from the beginning. Policy was formulated at national, regional and local levels and bound together in several policy processes. It was presented to citizens as a political subject. Noise was seen as a source of political conflict rather than just a technical planning issue. Local, regional and national governments came into positions of influence in this process, with an emphasis placed on the role of the Regional Government.

At the end of the 1990s, federalization resulted in what I call a distribution discourse. Literally, the concept of distributing aircraft movements was introduced into the policy process by a member of the Cantonal parliament. This politician was especially concerned at the large number of aircraft passing over his own constituency. When he became Minister of economic affairs for the Canton, and was thus responsible for airport policy, he returned to the idea of distribution. This time, however, distribution also meant increasing the number of flights. The liberalization of civil aviation had led to more lenient noise abatement policies in Switzerland and the Canton pleaded for new flight paths under the heading of 'distribution'. It became the central storyline because it could support different positions: sharing the burden of noise while increasing total noise emissions. Along with the first connotation came ideas of 'solidarity' and the democratic protection of minorities: the majority of the citizens of the Canton should not put the burden of noise on a small number of villages. As such, distribution confronted a central question of Swiss federal policy: how to reconcile the different layers of the polity.

Within the distribution discourse, aircraft noise was redefined as something that had to be distributed differently (and preferably fairly). Aircraft movements and flight paths became the defining characteristics of noise in the public debate. In contrast, in the Netherlands, annoyance is primarily defined in terms of an acoustic load that is present in particular areas. The Swiss distribution policy coincided with a regional consultation process, new legislation, a dispute with Germany, and the dissemination of information on possible flight paths by the airport. The proposal that aircraft movement might be distributed differently in the future alarmed many neighbouring communities and citizens. Within a year, dozens of new protest groups emerged, all directed at the protection of their own local community. Distribution alarmed citizens in quiet areas who believed that, in the future, they might be exposed to aircraft noise. Distribution did, however, present a window of opportunity for already exposed communities who argued for 'solidarity' and urged other communities to 'take their share' of the noise. In this context, aircraft noise was framed as a threat to local communities or *Heimat*.

The ensuing conflict even undermined existing consultation bodies. The regional consultation body of local communities reached a compromise on distribution, which the Canton largely disregarded. The Canton, for its part, appointed a similar body in 2003–2004 which comprised a round table of local communities. This initiative quickly failed due to the impossibility of determining which communities should participate. The distribution policy had alerted so many people and local politicians to the issue that about 200 different political bodies wanted to participate in the talks. The distribution discourse created conflicts that could not be solved by the institutions that had produced the distribution discourse in the first place.

In summary, it is apparent that through different policy processes in Amsterdam and Zurich, two extremely distinct problems arose from aircraft noise. In Amsterdam, it became a problem of ecological modernization and technical solutions, closely linked to sustainability while in Zurich noise turned into a political conflict about democracy.

NOISE PERCEPTIONS IN AMSTERDAM AND ZURICH

The focus on noise annoyance as the central criterion for limiting the growth of Amsterdam Schiphol only had a very limited effect in the sense that the airport kept on growing. Noise policy might have stimulated the use of more efficient engines and experimentation with air traffic control procedures that were designed to confer quieter takeoffs and landings and thus mitigate noise exposure to some degree. In Switzerland, annoyance policy hampered airport expansion in the sense that the ensuing political conflicts delayed policy-making severely. In both cases, market development and airport expansion was more important for the ecological impact of civil aviation than annoyance policy or environmental policy.

Noise annoyance policy did, however, achieve a cultural shift as aircraft sound increasingly came to be associated with noise pollution. Indeed, over time, the specific policy discourse about aircraft noise annoyance came to dominate citizens' perceptions of airport expansion. The following data are derived from a discourse analysis of interviews, letters and citizens' complaints in Amsterdam and Zurich (see Bröer, 2006, 2007 for details on data collection and coding), a q-methodological study (Kroesen & Bröer, 2009) and surveys (Kroesen, Molin, & van Wee, 2010, 2011; Kroesen & Schreckenberg, 2011). The cultural impact of noise policy on perception can be demonstrated in four ways: firstly, people explicitly refer to policies when they talk about annoyance. Secondly, people use the policy discourse; thirdly, people struggle against the policy discourse; and finally the discourse people use influences their levels of annoyance.

First of all, politics and policies are major issues when people talk about aircraft noise. Almost half of the arguments about aircraft noise include specific references to policy, policy-makers or politicians (about 1,000 arguments coded in Atlas.ti). When asked about what they hear based on their own experiences, Dutch people are particularly quick to refer to politics and what 'they do' in government as well as 'broken promises'. Noise, then, is not a question of decibels, but of a relation, of 'you expose me' in the words of Stallen (1999). Interestingly, when a Swiss man was asked if there were more, less or the same amount of aircraft movements, he replied that aircraft noise was not a problem for him, given the hard times the national carrier Swissair was experiencing (Zurich 4:34). This man was strongly in favour of the policy supporting Swissair and therefore did not 'problematize' the noise from the airport. Even when explicitly asked to judge the sound, people redirected their attention to the quality of the policy that governed it. Significantly, they related aircraft sound to politicians, the political influence of the industry and social movements. For them, experiencing aircraft noise means evaluating noise policy.

Secondly, what people say is often consistent with the dominant policy discourse. People regularly use the concepts, storylines and logic of the annoyance policy. In the case of Amsterdam, many people accepted the growth trend argument and the 'mainport' storyline that was invented in the policy process in order to describe the airport. Of more than 1,000 arguments about aircraft noise, 45 per cent coincide with the respective policy discourse. If we look at specific arguments, we can see, for example, that in Switzerland, noise is almost twice as often described as a problem of local living conditions as in the Netherlands. This relates to the distribution policy discourse that emphasizes the effect of noise on local communities, whereas the discourse in the Netherlands constructs noise as a general problem for the whole population.

In the Netherlands, to give another example, complaints are an important part of noise policies. Therefore, Dutch people talk about annoyance as a matter of complaints almost three times as frequently as Swiss citizens. More specifically, noise complaints show how a dominant discursive practice structures the intricacies of noise annoyance. The way people report annoyance to official bodies in the two respective cases follows the requirements of the agency. When people are asked to complain in terms of the number of times they were disturbed, as in the Netherlands, they provide this information to express their complaints. If people are asked to describe their specific local situations, as in Switzerland, this is how they frame their complaints.

In addition to the content of the complaints, the number of complaints also changes in response to emerging political opportunities. In the Netherlands, complaints have become more important in the mainport and environment discourse. The Dutch government redefined itself as being responsive to complaints and displayed interactive policy-making as a part of airport planning. Citizens were advised to raise concerns about different policy proposals. This advice turned complaints into a vote on planning policy and provided a way for citizens to express concerns about future noise levels. When policy offered a window of opportunity, the number of complaints went up. Noise annoyance policy brought forth massive complaints in the Netherlands. The agency at Zurich Kloten has not, with one exception, facilitated serial complaining, and the number of complaints has remained low.

Third, people struggle with the policy discourse. The intrusive power of a policy discourse shows when individuals or groups try to frame aviation and noise differently. Most criticism and many opposition activities conformed to the dominant discourse to some degree and added something to it. Thus, in the case of Amsterdam, critics either placed more emphasis on the environment or the economic side of the policy. In Switzerland, opposition centred on where and how to distribute noise annoyance and flight paths. There was hardly any doubt about the distribution framework as such.

The struggle with the dominant discourse goes hand in hand with certain argumentative features like 'ridiculing' or 'exaggeration'. A typical exaggeration would be to describe the whole sector and government as acting in conspiracy. Often, sharp criticism is coupled with feelings of powerlessness. In interviews, letters and complaints, people simultaneously try to exert influence and express negative expectations about their own power. This is not fatalism but a contradiction that is ingrained in the policy: citizens are called on to participate without clearly effecting participatory procedures.

Another example of dissonance is how citizens use the complaint procedure. Some citizens submit thousands of complaints per year. This is facilitated by easy-to-fill-in registration forms. These forms specifically ask people to register every single aircraft that causes annoyance. Citizens use these forms, and at the same time use the open free text section of the forms to criticize both policy and, more specifically, the complaints agency. Therefore, it is possible that a person will hand in a pile of complaints in the act of arguing against the practice of collecting piles of complaints. This dissonance is argumentatively resolved with sharp criticism and cynicism. It is worth noting that the complaints agency does not answer complaints and does not ask for 'qualitative' information. Additional comments are thus not strategic or necessary. Instead, they signify a struggle with the dominant discursive practice. Dissonant positions reproduce the policy discourse as they introduce strong opposition.

Fourth, the discourse people use influences their level of annoyance. The positions that citizens take towards the policy discourse impact on how annoying they consider the noise to be. Those who align with the dominant discourse (consonant relation) reveal average scores. When people oppose part of the policy discourse (dissonant relation), their scores strongly deviate from the population average. In the Netherlands for example, if one opposes the ecological modernization discourse because it is not 'ecological' enough, annoyance is significantly higher. If one stresses 'modernization' instead, annoyance is significantly lower (Kroesen & Bröer, 2009; Kroesen et al., 2011). It is important to remember that this is not a question of a singular opinion on an issue but the entirety of different opinions that we call discourse, frame or even subjectivity, which is demonstrably shaped by policy discourse.

PARTICIPATORY PRACTICES

Zurich Airport has always been subjected to direct democratic procedures. In the Netherlands, Schiphol has become part of interactive planning since the mid-1980s. Since then, large-scale facilities like airports have had to go through formal assessment and consultation procedures and policy-makers invite participation from a wide range of stakeholders. Complaint procedures, informational meetings, large societal discussions, research and even the financing of activist groups were all parts of an attempt to democratize planning.

However, as this chapter has demonstrated, participation is not without contradiction, especially in the Netherlands. For example, complaint procedures define noise annoyance as an individual matter, but the policy instruments that steer noise exposure – limits, runways and flight paths – are based on the understanding of annoyance as a phenomenon of the whole population. Here, psychological and epidemiological approaches contradict each other. Many of these contradictions are the result of 'layering' in path dependent policy-making: new policy is superimposed on an old policy rather than replacing it. Before concluding this chapter, a personal observation might highlight the intricacies of these participatory practices that are inserted into a centralist planning tradition.

Recently I was invited to speak at a meeting of the so-called Alders roundtable, named after its chairman and former minister of environmental affairs. The roundtable had been introduced in 2006, three years after the opening of Schiphol's newest runway. Originally devised as the solution to both capacity and noise problems, the runway proposals were confronted with opposition from communities that hitherto had not been involved. The national government tried to 'regionalize' the noise issue and involve communities more directly. The roundtable strongly resembled the existing committee for regional consultation but had a larger representation from the aviation sector. Central government asked the round table to come up with a policy to enable growth without further noise annoyance. Between 2006 and 2010, the roundtable did indeed come up with a policy proposal but, in the process, a number of critical resident groups left the roundtable.

Recently, the roundtable and the regional committee had to merge into a neighbourhood council. Once again, this council sought to 'involve citizens as much as possible', as it was explained in the text that accompanied my invitation. I was asked to write and speak about this objective at preparatory sessions. I replied by suggesting that it was less productive to invite me or more experts, and the council should direct its attention towards unrepresented critics and action groups concerned with similar issues. In response, I was informed that 'the interests of activists are recognized, but they have a role to play outside the council' (email from the regional committee, May 16, 2012). And: 'given the choices made in the past, these sessions are meant to improve the current way of participation' (email, May 22, 2012).

This anecdote exemplifies participatory practices around Schiphol Airport. The contradiction between legitimizing annoyance and continuing noise exposure is not resolved. The policy process implies that citizens are active co-producers of the policy, but also that this only applies within the framework of ecological modernization and in accordance with pre-given choices. Participation means involving those who are willing to work within the framework of national policy. This also applies to the position of experts.

POLICY RELEVANCE

Policy and mobilization towards sustainability need emblematic issues. In the 1980s, acid rain and dying trees served this purpose. Noise pollution, was and in part remains emblematic of the ecological footprint of civil aviation. However, if one wants to limit the growth of civil aviation to achieve sustainability or some other objective, noise annoyance policy might not offer the most productive point of departure. This is exemplified by the fact that earlier goals for CO_2 reduction at Schiphol have been sidelined by the fixation on noise pollution (as, very often, reducing noise means increasing CO_2 emissions).

Noise annoyance policy did 'localize' the negative side effects of aviation and provided an entry point for mobilization but, over time, this gave rise to a culture of annoyance in a very specific way. Action groups, lobbies, industry, science and politics grew accustomed to a certain definition of the noise problem. Annoyance became an urgent problem and a legitimate complaint. It seems plausible that noise annoyance policy itself made people more sensitive to noise and heightened the associated annoyance. Citizens were mobilized to express their concerns in manifold ways and yet, in numerous cases, the decision to expand air traffic had already been made. Annoyance policies often mobilize concerns without providing solutions. From the perspective employed here, people's concerns are moving targets that change when addressed. While this cannot be overcome, policies can be scrutinized for their potential impact on perceptions before they are implemented.

Does this mean that protests against noise or participation have no effects? No, but it emphasizes the point that the focus on annoyance and the ways that citizens participate are in themselves the outcomes of political decisions that are in favour of airport growth. And, on a psychological note, it is possible that, over time, citizens and experts who become involved in the noise issue will forget about larger agendas. This is sometimes expressed when anti-noise groups align with other groups such as conservationists or anti-capitalists. In other words, while participation and protest are vital for democratic politics, it is essential to question how the issue is defined and to scrutinize the existing framework in which participation, in whatever form, becomes possible. As it stands, participation is often framed in terms of growth, noise and pollution.

Finally, airport development is a major site of interaction between citizens and other 'stakeholders'. As there are no universally accepted criteria for defining or identifying stakeholders, every attempt at inclusion will produce new fissions. Antagonistic processes themselves are not problematic. It is the basic decision to accept the negative impact of civil aviation that is covered up in these processes and it is this that prevents more productive solutions. A moratorium on 'growth' for example, might create the space that is necessary for fundamentally rethinking aviation and sustainability in the future.

REFERENCES

Bailey, P. (1996). Breaking the sound barrier: A historian listens to noise. *Body and Society*, *2*(2), 49–66.

Bijsterveld, K. (2008). *Mechanical sound: Technology, culture and public problems of noise in the twentieth century*. Cambridge, MA: MIT Press.

Boons, F., van Buuren, A., & Teisman, G. (2010). Governance of sustainability at airports: Moving beyond the debate between growth and noise. *Natural Resources Forum*, *34*(4), 303–313.

Bröer, C. (2006). Beleid vormt overlast, hoe beleidsdiscoursen de beleving van geluid bepalen *(Policy annoyance, how policy discourses shape the experience of aircraft sound)*. Amsterdam: Aksant.

Bröer, C. (2007). Aircraft noise and risk politics. *Health Risk & Society*, *9*(2), 37–52.

Bröer, C., Wirth, K., Brink, M., & Schierz., C. (2005). Veränderung der lärmbelästigung im zeitlichen verlauf. *Umweltmedizin in Forschung Und Praxis*, *10*(2), 105–112.

Dierikx, M., & Bouwens, B. (1997). *Building castles of the air: Schiphol Amsterdam and the development of airport infrastructure in Europe, 1916–1996*. The Hague: SDU Publishers.

Eidgenössisches Expertencommission. (1963). *Lärmbekämpfung in der schweiz*. Bern, Switzerland: Eidgenössisches Justiz- und Polizeidepartement.

Fidell, S. (2003). The Schultz curve 25 years later: A research perspective. *Journal of the Acoustical Society of America*, *114*(6), 3007–3015.

FoE. (2012) *Aviation and climate change report*. Retrieved from http://www.foe.co.uk/resource/reports/aviation_climate_change.pdf. Accessed in November 2012.

Gössling, S., & Upham, P. (2009). *Climate change and aviation: Issues, challenges and solutions*. London: Earthscan.

Hajer, M. (1995). *The politics of environmental discourse, ecological modernization and the policy process*. Oxford: Claredon Press.

Heppenheimer, T. A. (1995). *Turbulent skies, the history of commercial aviation*. New York, NY: Wiley.

Huys, M. (2011). *Building castles in the (Dutch) air: Understanding the policy deadlink of Amsterdam Airport Schiphol 1989–2009*. Delft, the Netherlands: Eburon.

Job, R. F. S., Topple, A., Carter, N. L., Peploe, R., Taylor, R., & Morell, S. (1996). *Public reactions to changes in noise levels around Sydney airport*. Internoise 1996, Liverpool.

Kroesen, M. (2010). New opportunities for aircraft noise policy in the Netherlands: Air transport and operations. *Proceedings of the first international air transport and operations symposium 2010*, p. 253.

Kroesen, M., & Bröer, C. (2009). Policy discourse and people's internal frames on declared aircraft annoyance: An application of Q-methodology. *Journal of the Acoustical Society of America*, *126*(1), 195–207.

Kroesen, M., Molin, E. J. E., & van Wee, B. (2010). Determining the direction of causality between psychological factors and aircraft noise annoyance. *Noise & Health*, *12*(46), 17–25.

Kroesen, M., Molin, E. J. E., & van Wee, B. (2011). Policy, personal dispositions and the evaluation of aircraft noise. *Journal of Environmental Psychology*, *31*(2), 147–157.

Kroesen, M., & Schreckenberg, D. (2011). A measurement model for general noise reaction in response to aircraft noise. *Journal of the Acoustical Society of America*, *129*(1), 200–210.

Maris, E., Stallen, P. J., Vermunt, R., & Steensma, H. (2007). Noise within the social context: Annoyance reduction through fair procedures. *Journal of the Acoustical Society of America*, *121*(4), 2000–2010.

Schafer, R. M. (1994). *Soundscape, our sonic environment and the tuning of the world*. Rochester, UK: Destiny Books.

Schreckenberg, D., & Meis, M. (2007). Noise annoyance around an international airport planned to be extended. *INTER-NOISE and NOISE-CON congress and conference proceedings*, *2007*(4), 3351–3360.

Schultz, T. J. (1978). Synthesis on social surveys on noise annoyance. *Journal of the Acoustical Society of America*, *64*, 377–405.

Stallen, P. J. (1999). A theoretical framework for environmental noise annoyance. *Noise and Health*, *3*, 69–80.
Thomas, C., & Lever, M. (2003). Aircraft noise, community relations and stakeholder involvement. *Towards Sustainable Aviation*, 97–112.
Upham, P. (2003). *Towards sustainable aviation*. London: Earthscan.
van Eeten, M. (1999). *Dialogues of the deaf: Defining new agendas for environmental deadlocks*. Delft, the Netherlands: Eburon.
van Kempen, E. E. M. M., & van Kamp, I. (2005). *Annoyance from air traffic noise: Possible trends in exposure-response relationships*. Bilthoven, the Netherlands: National Institute for Public Health and the Environment (RIVM).
Walker, S., & Cook, M. (2009). The contested concept of sustainable aviation. *Sustainable Development*, *17*(6), 378–390.

CHAPTER 8

COMPETITION, INTEGRATION, SUBSTITUTION: MYTHS AND REALITIES CONCERNING THE RELATIONSHIP BETWEEN HIGH-SPEED RAIL AND AIR TRANSPORT IN EUROPE

Frédéric Dobruszkes and Moshe Givoni

ABSTRACT

Purpose – *This chapter provides a critical discussion of air to rail mode substitution. Environmental impacts, intermodal competition and integra tion are considered, examining advantages and disadvantages as well as opportunities and constraints.*

Originality – *Both operation and life-cycle analysis perspectives show that high-speed rail (HSR) is much 'greener' than air transport (per seat-km or per passenger-km) provided that the former achieves high load factors and the latter lower load factors and that freed runway capacity is not reused. HSR travel time is its main competitive advantage*

Sustainable Aviation Futures
Transport and Sustainability, Volume 4, 175–197

ISSN: 2044-9941/doi:10.1108/S2044-9941(2013)0000004008

against air transport, and a 600-km flight is arguably the current limit for robust intermodal effects.

Findings – *The potential for air–HSR integration at the airport relies on various service, business and technical constraints. Even when it is successful, its environmental benefit appears to be marginal, if not negative, if airport capacity is reused for longer flights. In the current context, such integration appears more like a business opportunity for airlines, airports and train operators rather than a sustainable option. Yet the environmental benefit of integration may be larger within potential integrated transport policies.*

Keywords: High-speed rail; mode substitution; intermodal competition; air–HSR competition; air–HSR integration; air–rail integration

FROM REVENUES TO GREENER MOBILITIES: A BRIEF HISTORY OF HSR IN EUROPE

When the first European high-speed rail (HSR) service was launched in 1981 between Paris and Lyons, nobody had in mind its potential to achieve 'green' or 'greener' mobilities. Intermodal competition, declining traffic and concerns about their financial state led the French national railway company (rather than the State) to start research on high speeds in the 1960s and then to develop their HSR with the support of rolling stock manufacturers (Fourniau, 1989; Klein, 2001; Powell, 1997). Clear evidence comes from French President François Mitterrand's inaugural speech in 1981: he never tackled environmental issues; he only quoted energy savings, which at that time were seen as a geostrategic concern.[1]

Since then, things have really changed and successive European White Papers expect large expansions and modal shifts to HSR. The EC's (2011) White Paper, in the section on 'Ten Goals for a competitive and resource efficient transport system: benchmarks for achieving the 60% GHG emission reduction target', aims to:

> By 2050, complete a European high-speed rail network. Triple the length of the existing high-speed rail network by 2030 and maintain a dense railway network in all Member States. By 2050 the majority of medium-distance passenger transport should go by rail. (EC, 2011, p. 9)

The reasons for this are evident. HSR is renowned for consuming less energy and emitting fewer greenhouse gases (GHGs) and pollutants per passenger-km than aircraft and cars. Some routes have experienced modal change from air to rail following the introduction of HSR. Such mode substitution can take two forms: intermodal competition or intermodal integration. The former means that services compete head-on; operators do their best to get as high a market share as possible. The latter supposes that each mode is used on its best distance range and may involve various levels of cooperation between operators.

The aim of this chapter is to provide a critical discussion of air to rail mode substitution. Its environmental benefits are first reviewed. Then, competition and integration are explored along two lines: common beliefs based on the available evidence are examined and the balance between advantages and disadvantages as well as between opportunities and constraints is discussed.

THE ENVIRONMENTAL BENEFIT OF MODE SUBSTITUTION TO HSR

Table 1 shows a comparison of the environmental damage caused by a flight and an HSR journey between London and Paris with respect to local air pollution and climate change impacts (see Givoni, 2007 for full details of the methodology and assumptions used). The advantage of the HSR over the aircraft is clear, especially considering both the local air pollution and climate change impacts. Although the analysis was carried out for a specific route and a specific aircraft (an Airbus A320-200), the magnitude of the difference between the modes means that HSR would retain its advantage even when the assumptions are changed, for example, on longer routes up to distances where the two modes still compete, and similarly if

Table 1. The Environmental Damage from Aircraft and HSR Journeys on the London–Paris Route (Euro/seat).

	Local Air pollution (LAP)	Climate Change (CC)	LAP + CC
Aircraft	0.82	2.03	2.85
HSR	0.52	0.29	0.81

Source: Based on Givoni (2007).

considering the latest generation of passenger aircraft. HSR's environmental performance can be even better than that reported in Table 1 depending on the extent to which renewable energy is used to generate the electricity that is required to power the HSR. In France, where electricity is mainly from nuclear energy, the environmental impact of HSR is lower, notwithstanding concerns on nuclear waste and risks.

With respect to other environmental impacts from aircraft and HSR operation, it is more difficult to determine what the damage is and which mode is responsible for more of it. Such impacts include noise, land take, and loss of biodiversity.

Notwithstanding the above, analysis by Chester and Horvath (2010) on the California HSR demonstrates how its environmental performance, compared to other modes of transport including aircraft, can vary significantly depending on the load factor achieved. On the London–Paris route, both modes achieve a relatively high load factor. However, when HSR achieves a low load factor and the aircraft a high load factor, the latter results in significantly less energy consumption per passenger-km. Given the capacity of high-speed trains (in Europe, more than 300 seats for the shortest ones and up to 750 seats, compared to 100–200 seats for narrow body aircraft commonly used on short-haul routes), it is much more difficult to achieve a high load factor.

The analysis by Chester and Horvath (2010) also demonstrates the importance of looking at phases other than operation in the life-cycle analysis of both modes. These phases (infrastructure and vehicle-related environmental impact during construction/production, maintenance and disposal) can be responsible for significant environmental impacts, although much less than the impact from the operational phase. Comparing between HSR and aircraft in the California case (the planned HSR) shows that under a high low load factor scenario the HSR impact on the environment is lower. Yet, when a low load factor is assumed the environmental impact from HSR is much higher than that of the aircraft.

Since HSR has a considerable environmental advantage per passenger-km over aircraft if high traffic density and load factors are met, it is expected that mode substitution from air to rail and policies to promote such mode substitution will result in environmental gains. While a valid assumption, there are many reasons why this might not be the case. The net environmental effect of mode substitution depends very much on what happens with the freed runway capacity at the airport. If this capacity is used to accommodate more flights, there would be no environmental gains from mode substitution. Such gains also depend on other characteristics of the route, like the extent to which such mode substitution is expected. For the London–Manchester route, Miyoshi and Givoni (2013) estimate that the amount of carbon mitigated by the introduction of HSR on the route will be small. Overall CO_2 reduction will be in the order of 100,000 tonnes per annum in 2033, which is less than 0.1% of the total UK domestic transport emissions in 2007. This is mainly because modal shift from air and road is relatively small on the London–Manchester route and because the majority of electricity used to power UK railways relies heavily on non-renewable and non-nuclear energy sources.

AIR–HSR COMPETITION: MYTHS AND REALITIES

HSR requires less energy and generates fewer GHGs and pollutants per passenger-km than air transport. This section analyses the extent to which HSR is able to compete with air services using the available evidence.

Mode Substitution: What Do We Know?

Literature reviews on intermodal effects following the introduction of new HSR services show that there is much more published work on the ex-ante expected intermodal competition than on ex-post observed change (Capon, Longo, & Santorini, 2003; Givoni & Dobruszkes, 2013; Wardman, Bristow, Toner, & Tweddle, 2002). Furthermore, researchers usually focus on the demand (i.e. the passengers) but not on the supply of services (e.g. the number of flights, available seats or seat-km) and consider a limited range of routes (see, e.g., Behrens & Pels, 2009; Campos & Gagnepain, 2009; Cheng, 2010; Cascetta, Papola, Pagliara, & Marzano, 2011; Dobruszkes, 2011; EC, 1998; Jiménez & Betancor, 2012; Lee, Yoo, & Jung, 2012). Very often, ex-post studies conducted or commissioned by transport companies are usually unavailable, owing to their strategic value in a competitive context.

Assessing the intermodal impacts of HSR raises the issue of the data used. It is common to plot HSR market shares versus HSR travel times (Fig. 1). Indeed, when considering HSR and air transport only, a good relationship between the two variables is found.[2] HSR unsurprisingly reaches higher market shares with lower travel times – more than 50% up to 3.5 hours. This is consistent with other works that highlight travel time

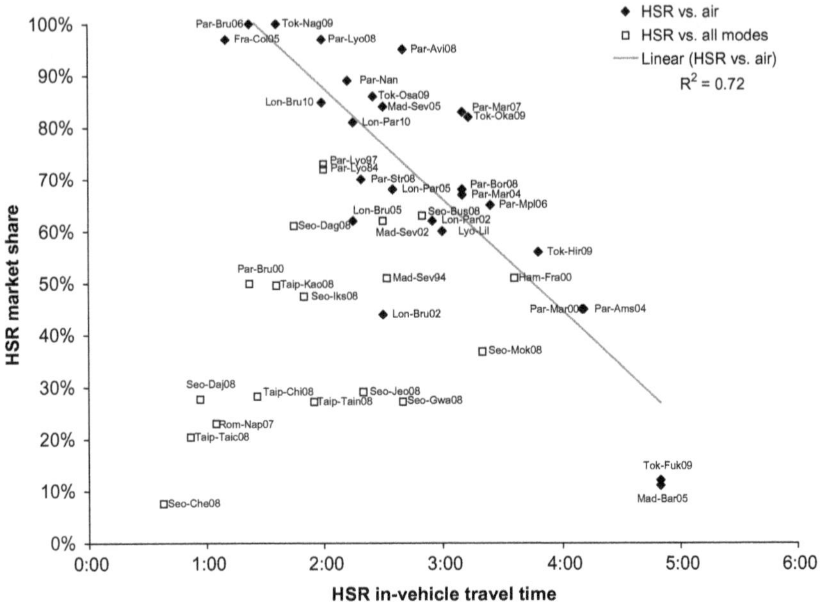

Fig. 1. The Traditional Plot: HSR Market Share as a Function of HSR Travel Time. Figure Compiled by the Authors Using Available Ex-post Figures.

as a main factor in intermodal competition (e.g. Cascetta et al., 2011; Coto-Millán, Inglada, & Rey, 2007; Dobruszkes, Dehon, & Givoni, 2012; EC, 1998; Steer Davies Gleave, 2006; Vickerman, 1997). Market share dispersion for a given HSR travel time nevertheless suggests that other factors are also at work, such as fares, access and egress times, frequencies, schedules, airline connections, comfort and reliability. For example, Chang and Lee (2008) found that price (38%) and poor HSR station accessibility (32%) were the main reasons why travellers did not use the Korean HSR when users of all modes were surveyed on the Seoul–Busan and Seoul–Mokpo corridors.

Other authors have also reported market shares before/after HSR. However, this raises the issue of induced traffic. HSR passengers may come from both modal substitution (passengers taken from other modes) and induced traffic (new trips made). Although assessing induced traffic is notoriously difficult (see Bonsall, 1996, and Mokhtarian, Samaniego, Shumway, & Willits, 2002), surveys made in the aftermath of HSR launches in both Asia and Europe suggest that induced traffic ranges from

6% to 37% of HSR ridership (Givoni & Dobruszkes, 2013). This involves an increase in the total market size. The meaning of relative market shares thus changes from time to time, along with the market size. This becomes evident with the cross-channel market, where HSR was launched in 1994. Considering only airlines and HSR, Table 2 shows that airlines experienced a large fall in their market share: 81% and 85% on the London–Paris and London–Brussels routes, respectively. However, their loss of passengers was not as substantial (56% and 58%, respectively). This is largely due to the growth of the combined air–rail market over the same period, which can be considered as induced demand.

Figures have also been provided on HSR passengers as a percentage of passengers' mode of origin. Depending on the route, HSR ridership may be mostly made up of former airline passengers (e.g. 42% between Madrid and Seville (EC, 1998)) or traditional rail passengers (e.g. 69% between Rome and Naples (Cascetta et al., 2011)). However, such figures say nothing about the level of remaining airline passengers or services. This is why it is better to consider the change in ridership and/or services in absolute figures. Doing so[3] allows several conclusions to be drawn, assuming that

Table 2. Airline and HSR Cross-Channel Passengers (in thousands).

	1993	2010	2010/1993
London–Paris			
Airlines	3,665	1,626	−56%
	100%	19%	−81%
HSR	0	6.728	
	0%	81%	
Sum	3,665	8,354	+128%
London–Brussels			
Airlines	1,160	490	−58%
	100%	15%	−85%
HSR	0	2,801	
	0%	85%	
Sum	1,160	3,291	+184%

Sources: Brussels Airport, CAA, DGAC/DTA/SDE and Eurotunnel.

most, even if not all, of the dynamics are due to a shift to the HSR. First, HSR tends to have more impact on air transport than on other modes. However, when travel time is longer, HSR may have more impact on traditional rail services than on air services (as the Madrid–Seville case shows). Actually, this may be due to monopolistic rail companies forcing their passengers to switch to HSR by removing traditional services. Second, it is clear than mode substitution to HSR decreases when travel time increases. With lower travel times, HSR is in a position to cannibalise virtually all airline markets. Third, the impact on demand can be higher than on services, as illustrated by the Madrid–Zaragoza, Madrid–Seville or Seoul–Busan cases. This suggests that the environmental benefit can be lower than as suggested by the trend in ridership, since the emissions come from the aircraft and not from the passengers.

Finally, one should be careful with before/after figures. Usually, 'after' means only a few years after the introduction of HSR. However, the terms of the competition may later change. On the one hand, airlines' hubbing strategies may involve preserving or (re)expanding air services despite competition from HSR. Between Paris and Brussels, for example, HSR led to a substantial decrease in air services, and even to their temporary suspension (Fig. 2). Then, Brussels Airlines, the main airline serving Brussels airport, decided to recommence one and then two daily flights between the two capitals in an attempt to capture French travellers going to/from destinations in Africa, a continent well served by the airline. Despite the low market share, the number of flights in 2012 represented 35% of the 1991 level.[4] On the other hand, when HSR travel times are not really competitive, airlines can expand. The German flag carrier Lufthansa decreased its provision of flights on the Cologne–Munich route (a journey with an average HSR travel time of 4 hours 37 minutes on an incomplete high-speed line). This enabled low cost airlines (LCAs) Germanwings and DBA and hybrid carrier Air Berlin to enter this market,[5] competing with Lufthansa and HSR through lower prices. Thus, HSR did not prevent an increase in the provision of air services on this route (Fig. 2).

Whether the LCAs are a threat to HSR is debatable (Rothengatter, 2011). Behrens and Pels (2009) did not find an LCA effect on the London–Paris route in 2003. However, in Germany, where HSR travel times are often more than 3 hours given the geography of high-speed lines (HSLs) compared with the urban system, it is easier for airlines to compete with HSR. Considering 84 domestic routes, Friederiszick, Gantumur, Jayaraman, Röller, and Weinmann (2009) find that low-cost airlines' entry into the German market resulted in a decrease in rail passengers of at least

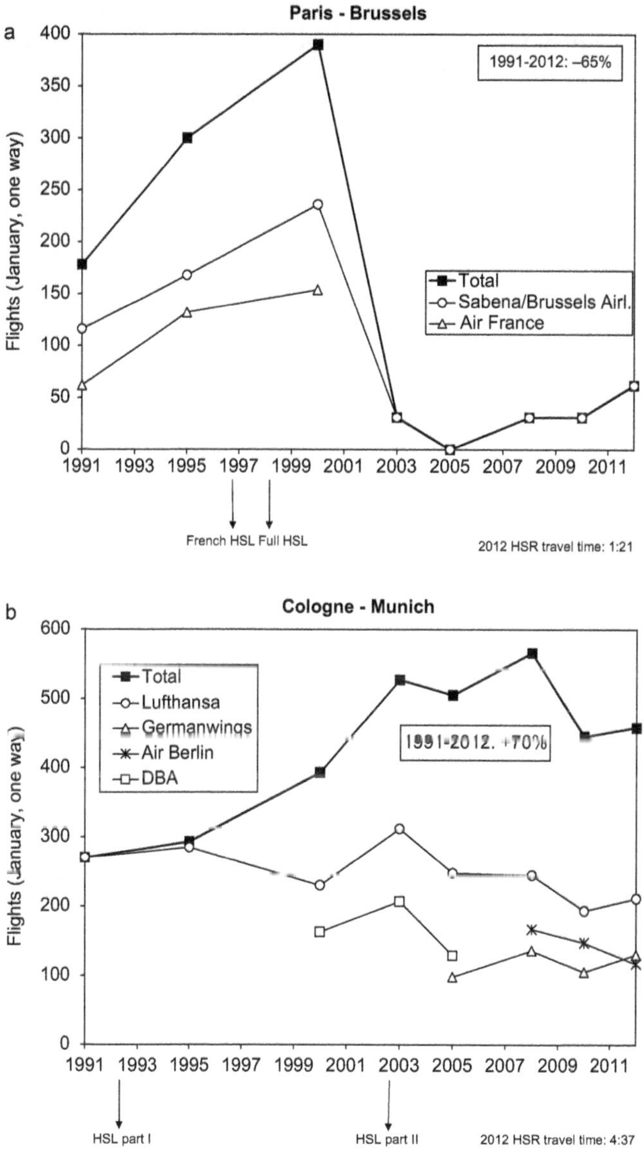

Fig. 2. Examples of Air Services Time Series.

7% in second-class and 18% in first-class travellers. The effect of LCAs on demand for rail was stronger on longer routes. In addition, a survey of four German routes reported by Friebel and Niffka (2009) indicates that 2.8–33.3% of LCAs' passengers would have chosen rail if LCA had not been available (if rail travel time were more than 10 hours or less than 4 hours, respectively).

In summary, HSR travel time appears to be a main factor in the ability of HSR to decrease the level of air demand and services, although one should be careful considering long time series and bear in mind that the LCAs are arguably a threat on longer routes.

The Geo-Economic Constraint: Traffic Density vs. Infrastructure Costs

HSR travel time (and thus HSR attractiveness) relies on both inter-city distances and the availability of HSLs. Commercial speed is subject to the number of stops, but with high-speed trains running at 300 kph, a 250 kph commercial speed can be assumed. HSR appears to be competitive up to around 3 hours (approximately 750 km). However, aircraft can fly more direct routes than HSR.[6] Assuming the latter experiences 25% longer routes on average, a 600 km flight is arguably the current limit for robust intermodal effects due to HSR.

How large is the potential for air–HSR substitution? The answer is linked to a mix of market, economic and geographical factors. The short-haul market represents a significant part of worldwide regular air services. In January 2012, 39% of flights and 28% of seats were for services of less than 600 km.[7] However, building HSLs is rather expensive.[8] Considering 45 HSR projects worldwide, Campos and de Rus (2009) find that infrastructure costs range from 6 to 45 million euros/km, with an average of 17.5 million (2005 values) plus planning and land costs. Thus, from a cost–benefit analysis perspective, only HSLs with high-density traffic can be socially profitable. For example, de Rus and Nombela (2007) estimate that at least 8–10 million annual passengers are required when opening a 500-km HSL. This threshold and underlying assumptions are open to criticism.[9] Eight million passengers correspond to around 22,000 daily passengers or 29 daily 750-seat HSR services. Few existing HSLs support such flows. Nevertheless, these figures highlight that traffic density is strategic. Givoni, Dobruszkes, and Lugo (2012) made an attempt to assess the potential for modal substitution based on air traffic density. The methodology still needs improvement, but they discovered that the share of high-density airline routes is much lower than the aforementioned share of short-distance services. Even with a smaller threshold of around 2 million airline seats per year and a more optimistic airline distance of 750 km, only 1.7% of the worldwide flights (or 0.6% of seat-km) could justify new HSLs. The fact that routes can be merged within the same corridor and thus on the same HSL is not considered here, so the potential is arguably higher. Irrespective, it can be concluded that if HSR can lead to dramatic decreases in air services on selected routes, the global potential for such modal substitution is rather limited.

Of course, these constraints are relevant within a free-modal-choice context. Let us consider which strong pressures would apply to the airline market. They may come either from the energy-side – because of the supposed peak oil and the current lack of fully substitutable fuel – or from governmental regulations concerned with local or global environmental impacts. Such constraints may strongly influence fares and/or the freedom of airlines. Either the demand would decrease or it would be switched to other modes including rail. Then, passenger density on many short-haul routes would increase and would justify HSLs, which are currently alleged to be socially unprofitable.

AIR-HSR INTEGRATION: A CRITICAL ANALYSIS

The previous section considered HSR as a means to compete with air services, especially through services from city-centre to city-centre. However, various HSR services now also serve airports, allowing integration between air/rail and airlines/train operating companies (TOCs).

Many traditional airlines have built hub-and-spoke networks, which are the combination of spatial concentration (star-shaped networks) and temporal concentration (timetable optimisation) (Burghouwt & de Wit, 2005). The airlines can thus operate a limited network and achieve economies of density to lower their operating costs (Brueckner & Spiller, 1994). Timetable optimisation should limit connecting times and transfers disbenefits. Adding one spoke in the network means an exponential increase of transfer opportunities. Hubs are especially useful to help airlines to fill their aircrafts, either on low-density regional routes or on long-haul markets.[10] In this context, HSR services can play the role of short-haul feeders, especially up to 600-km or 1-hour flights (Givoni & Banister, 2006).

HSR services serving airports do not necessarily involve airline–TOC integration. Integration depends on commercial agreements and on practical arrangements between the partners, as suggested by Table 3. Passengers

Table 3. A Taxonomy of Air–HSR Integration.

Integration	Type	Product	Places and Companies	Passenger Perspective
None	HSR services simply serving an airport	–	HSR at Paris CDG vs. Easyjet or American Airlines	Separate tickets must be purchased. Passengers assume all risks in case of delays or disruptions
Low	Interlining agreements	Rail&Fly	From/to any German (plus Basle) airports and any German railway station 77 airlines including Lufthansa and 77 tour operators including TUI + DB Bahn	Train tickets received from the airline (may be subject to an extra fee depending on the airline and the class of travel)
Medium	Code-share agreements	TGVAir	Paris CDG airport to/from 20 cities 8 airlines including Air France + SNCF	One booking; one ticket; pre-check-in at the railway station but no luggage handling
High	Joint-venture	AIRail	Frankfurt to/from Cologne, Bonn and Stuttgart Lufthansa + DB Bahn	Optimised train schedules. Online check-in or at the railway station. Luggage handled between Frankfurt AIRail Terminal and aircraft

Source: Based on Chiambaretto and Decker (2012) and companies' websites.

may take advantage of HSR rather than airline feeder services if they have better access to HSR stations than to airports or if the HSR frequency is higher than the airline alternative.

Since airline–TOC integration was successful at some airports, policy-makers have been inclined to recommend it as a general goal. However, its potential effects are subject to various market and technical factors (Chi & Crozet, 2004; Dobruszkes, 2001; Givoni & Banister, 2006). This section compares the Paris and Brussels cases to illustrate this.

Air Markets

Paris and Brussels differ in terms of air services. Paris is notably served by Paris Charles de Gaulle (CDG), one of the main airports in Europe and worldwide. Air France and its partners use it as a global hub. In early 2012, the airport was connected to 219 international cities, 138 of which are outside Europe. Such a wide range of connections involves a large catchment area and high proportion of connecting passengers. The closest global airport is London Heathrow, which can only be easily reached by air. Any other French airport is less well served and, symptomatically, only Nice also has air services to New York.

Brussels is mainly served by Brussels Airport, a medium-sized facility which nevertheless offers good connectivity (143 international routes) given the city's size. Services are mainly embedded in the European market but also include a significant number to Africa plus some more traditional long-haul destinations in North America and in Asia. However, Brussels is surrounded by the four biggest European airports: London Heathrow, Paris CDG, Frankfurt and Amsterdam. Furthermore, corresponding regions are also served by additional airports that mainly add European connections (Table 4, column A). As a result, considering the close regions that might be linked to Brussels Airport by HSR, it is clear that Brussels has fewer niche markets to offer than surrounding regions with their respective airports (Table 4, column B). Furthermore, Brussels' niche connections are small markets. For example, Brussels offers exclusive air links to Accra, Banjul, Kigali and Aqaba that cannot be found in Paris, but it is unlikely that the demand between Paris and these cities would justify high-speed trains to Brussels Airport. The low level of demand would involve few HSR services, which would make any train or flight delays or disruptions very problematic, as no robust alternative could be offered. In sum, HSR would be more useful to help Brussels' or Belgium's inhabitants

Table 4. Brussels Airport and Its Competitors.

Competitors	International Destinations (A)		Niches at Brussels (B)		Niches Against Brussels (C)	
	All	Non-European	All	Non-European	All	Non-European
Brussels	143	68	–	–	–	–
Amsterdam + Rotterdam	226	119	32	25	115	76
Frankfurt + Cologne + Düsseldorf + Dortmund	257	160	37	22	151	114
London (five airports)	319	145	35	24	211	101
Paris (CDG + Orly)+Lille	252	158	32	15	141	105

Source: Authors' computations based on OAG data (January 2012).

to access additional air links from airports abroad (Table 4, column C) than to convey passengers from France, Germany, the Netherlands or the United Kingdom to Brussels Airport.

Technical Constraints

The geography of railways is another difference between Paris and Brussels. Although French HSLs were mainly designed to link Paris through its terminal stations to other major French towns, all radial HSLs have been progressively interconnected since 1994 thanks to a ring railway made up of a new HSL in the east and an old traditional line in the south. This so-called 'Interconnexion' HSL directly serves Paris CDG airport (Fig. 3), with an HSR station embedded in its main terminal. This route involves virtually all inter-regional HSR services linking Northern France and Belgium to other French regions at Paris CDG airport, which is connected to 73 cities including Brussels.

In contrast, Brussels Airport is located away from HSR routes that link Brussels to both the Netherlands and Germany (Fig. 4). Since the 1950s, Brussels Airport has been served by trains arriving in a small terminal station. Airport expansion and rebuilding during the 1990s included a new railway station, again as a cul-de-sac. In the 2000s, it was decided to open up the terminal station thanks to new lines allowing direct links to/from the north and the east of the country (the partially completed 'diabolo project'). Once achieved, this would theoretically allow HSR services going to/from the Netherlands or Germany to directly serve Brussels Airport.

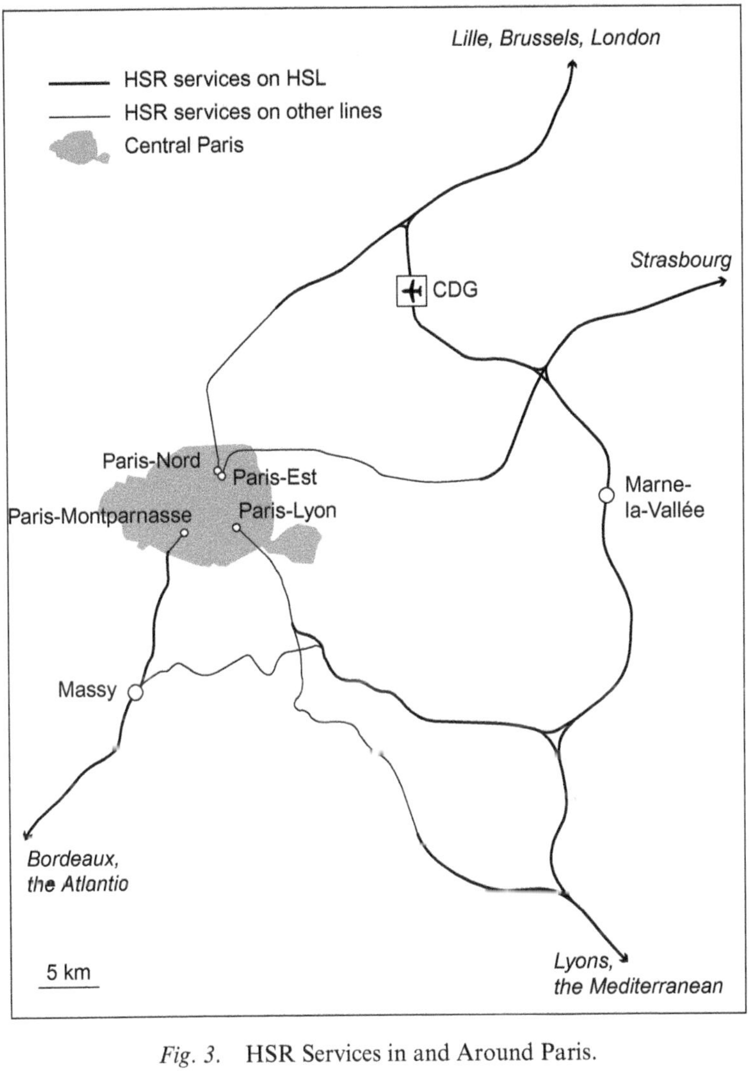

Fig. 3. HSR Services in and Around Paris.

However, there are various reasons for not rerouting HSR via Brussels Airport. First of all, the market for HSR–air integration is arguably limited. Second, as Fig. 4 shows, serving Brussels Airport will involve a significant detour and cost time (about 15–20 minutes) for those passengers travelling between the Netherlands or Germany and Brussels or Paris and

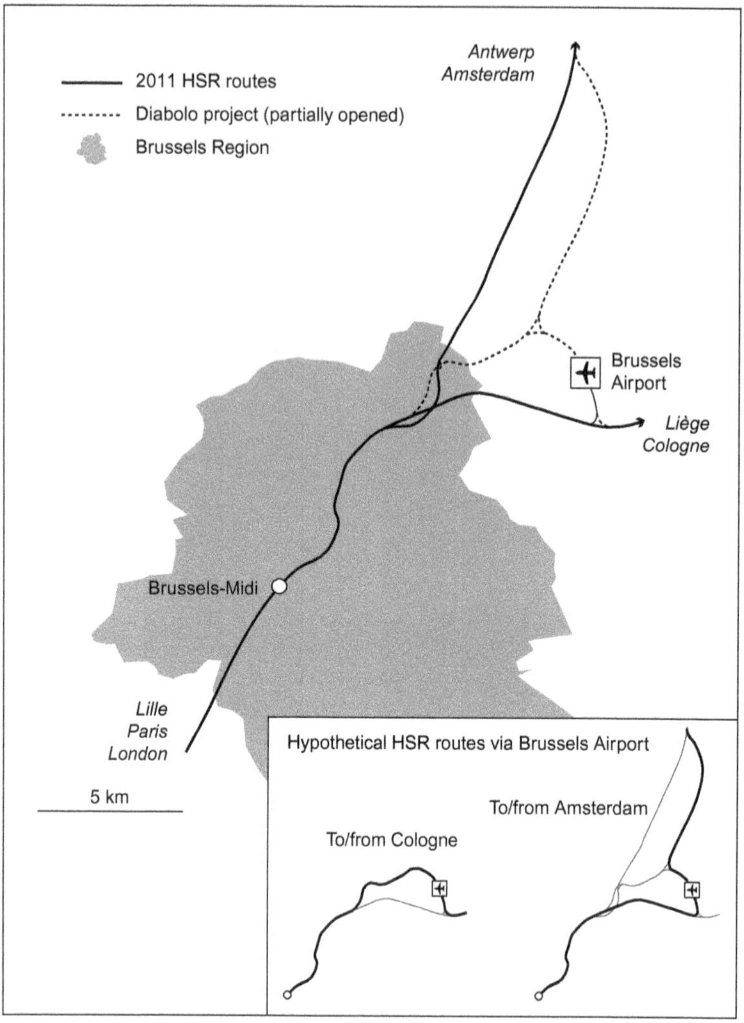

Fig. 4. HSR Services in the Brussels Area (valid early 2012).

beyond, i.e. the main market. In other words, serving Brussels Airport by HSR would make HSR less competitive in the main markets. Time lost may result in lost passengers (and thus revenues) and increased HSR fleet (and thus costs) to maintain the same level of frequencies. Finally, with only three railway tracks at Brussels Airport station, it is unclear whether adding HSR services would be compatible with a significant level of regional and national rail services.

Alternatively, additional HSR services coming from Paris and ending at Brussels-Midi station could be extended to Brussels Airport. Again, there is concern about limited demand. This involves adding a non-stop train on overloaded, traditional (i.e. non-HSR) tracks. The current daily service already involves the release of one return regional service from Brussels to the airport and one return journey from Brussels to Paris.

Finally, one could also consider linking Amsterdam and Cologne via Brussels Airport. However, this would involve not going on to Paris, not serving Brussels city-centre and missing the large range of rail services and urban transport available at Brussels-Midi station.

In sum, both air services patterns and railways patterns play against HSR services at Brussels Airport. Such services can only be set up to the detriment of main HSR services and to arguably more useful traditional rail services to the airport.

As Table 5 suggests, the social benefits of aircraft–HSR substitution seem limited. Furthermore, as stated in the second section, each case is unique. In a congested, global airport like Frankfurt, there is no doubt that HSR services can help airlines to release short-haul services, thus freeing slots for longer flights. This is good for both airlines and airports but certainly not for the environment, as airline emissions increase with distances flown.[11]

In an uncongested global airport like Paris CDG, HSR has helped the airport to increase its catchment area or, at least, to offer nationwide, seamless access/egress journeys to/from the airport by bypassing central Paris. However, in 2011, HSR delivered about 5% of Paris CDG passengers (Chiambaretto, Baudelaire, & Lavril, 2012). The environmental gain is thus arguably marginal compared to aviation impacts. In addition, HSR may have made flying easier for those travellers leaving near the HSR stations and thus more attractive.

Table 5. Comparing Emissions from Short- and Long-Haul Flights.

Avoided or Ended Flights	Corresponding GHG Emissions[a]	Selected Connecting Long-Haul Flights	Corresponding GHG Emissions[a]
Stuttgart–Frankfurt	60 kg	Frankfurt–Tokyo	3,410 kg
Marseilles–Paris	200 kg	Paris–New York	1,950 kg
Paris–Brussels	60 kg	Brussels–Kinshasa	2,020 kg

[a]In equivalent-CO_2 per passenger, one way, according to www.atmosfair.de.

At Brussels, the daily HSR service launched in 2011 to/from Paris is the result of an agreement between Thalys (the TOC), Brussels Airlines (the main airline serving Brussels Airport), Jet Airways (an Indian airline linking India and North America via its Brussels 'hourglass' hub) and Brussels Airport (a 75% privatised airport). Their shared press release[12] stated that the aim was to expand their business to the French market. Symptomatically, most long-haul connections highlighted in this document also exist from Paris CDG airport. Air–HSR integration may lead some long-haul passengers to switch from air services to HSR between Paris and Brussels. Again, the environmental gain would be marginal compared to the emissions generated by the long-haul segment of the journey. Other passengers to/from Paris might be attracted to Brussels Airport if fares were more attractive. Then, a local or regional access/egress journey to/from Paris CDG airport would be replaced with a longer HSR journey to Brussels Airport. Even with low energy consumption, the 328-km HSR journey would involve some additional emissions. In the end, air–HSR integration appears rather as a means to preserve or to expand the airline business.

This discussion has important policy implications. From an environmental perspective, it suggests that air–HSR integration should be promoted only if HSR would really secure a reduction in air services. However, this is difficult to foresee within the current context where meeting growing demand for air transport is the policy. First, air transport liberalisations make it easier to reuse freed runway capacity. Second, many airports, both public and private, have moved from technical- to business-oriented management. Privatised airports are of course profit-led and seek to maximise operations to make profits. Public airports have also often become more entrepreneurial, because local or regional authorities have themselves largely moved from traditional spatial planning to entrepreneurial, regional development (Harvey, 1989). Competing with each other for international attractiveness, public authorities notably aim to secure good long-distance accessibility. In this context, airports and air services are often seen as making the areas serviced appealing. Public authorities thus tend to facilitate airport extensions and air service developments.

These private and public objectives do not favour severe restrictions on air transport in the name of the environment. New European regulations may be relevant, for example, to ban airport extension in the case of significant HSR services. Yet until now, the EU has been rather schizophrenic in terms of air transport versus the environment. On the one hand, the EU promoted air transport liberalisation, which led to its massive expansion in terms of routes operated and volumes supplied (Dobruszkes, 2008).

Various EU bodies have officially welcomed this outcome. For example, the European Commission has once commented on the boom of the low-cost airline services as follows:

> The Commission welcomes this development and appreciates the contribution made by these companies to the general reduction in the price of air travel in Europe, the wider range of services available, and the accessibility of air travel to a wider public. (EC, 2005, p. 4)

On the other hand, the EU has long claimed that it wants to promote more sustainable mobilities and mode substitution from air to rail (EC, 2011). Since 2012, the EU has included all international aviation serving it in its emission trading scheme, but this is expected to have only very limited effects on air transport emissions without a more restrictive system design (Vespermann & Wald, 2011). In sum, would it be possible to achieve greener mobilities while saving free modal choice (and ultimately the airline business)?

CONCLUSIONS

The many years that HSR has been operating in Europe alongside air transport services allow a reflection on issues relating to competition, integration and substitution between the modes and an opportunity to distinguish between the myths and realities relating to the relationships between these modes of transport.

HSR can successfully compete with air services; therefore, it is expected that mode substitution from air to rail would take place when new HSR services are launched. The level of expectation for mode substitution, however, does not necessarily align with empirical evidence on mode substitution (across all HSR routes) which is more modest (Givoni & Dobruszkes, 2013). From this misconception about the likely level of mode substitution, several other myths arise. There is no doubt that per seat-km and even passenger-km HSR is greener than the aircraft. It is natural therefore to expect HSR to play a role in addressing the environmental problem of air transport, but expecting HSR to solve that problem alone leads to overexpectations from policymakers. Similarly, and for different reasons, it is unrealistic to expect air–HSR substitution to relieve congestion at major airports. These myths, as we refer to them, lead in turn to other misconceptions about the effect of air–HSR competition and integration and to ill-planned EU and national policies related to both air and rail transport.

The reality with respect to the relationships between HSR and air transport is a mixed picture, which varies significantly between routes, airports and general EU, national, regional and local circumstances. HSR can and should play an important role in the future of air transport. It could contribute to better management of air transport services in the face of growing environmental challenges, but only as part of a package of solutions that will also address other elements of air transport management, for example, air traffic control and airport congestion.

In conclusion, HSR is a very important mode of transport that needs to be considered as a strategic long-distance surface mode of transport situated between long-haul aircraft services (and short-haul flights over water) and other medium-distance surface and transport modes, like conventional rail and car. While competition between air and rail modes is very much a reality resulting mainly from the institutional division between air transport, rail transport, and other modes and, consequently, the disparate planning of the respective networks and infrastructure, there must be a move towards integration of these modes and their planning (Givoni & Banister, 2010). Only then can the conditions be set for the expectations from air and rail substitution to be realized and result in real benefits and not remain a myth policymakers just believe in.

NOTES

1. Address by President Mitterrand, 22 September 1981, obtained from the François Mitterrand Institute.
2. By contrast, when considering all modes, including cars and coaches, HSR market share is not correlated with travel time, notably because road transport is often cheaper and more flexible. Analysing road modes, however, is beyond the scope of this chapter.
3. Ongoing DATE project results have not been published yet.
4. And around 50% in terms of seats supplied.
5. Air Berlin is considered hybrid, because its business model is between a traditional airline and an LCA airline. DBA was bought by Air Berlin in 2006 and then dissolved.
6. For example, distance as the crow flies (vs. HSR distance) are 481 km (621 km) between Madrid and Barcelona, 398 km (524 km) between Paris and Amsterdam and 435 km (621 km) between Munich and Cologne (according to OAG and Thomas Cook).
7. Authors' computations based on OAG datasets.
8. Rolling stock is rather cheap compared to planes.

9. In particular, only time-savings and generated traffic are considered as benefits. The authors recognise that congestion and the need for additional capacity may decrease the threshold found. The environmental benefits of mode substitution are not considered.

10. Francis, Dennis, Ison, and Humphreys (2007) highlight that even between two world cities like London and Chicago, 60% of the American Airlines' passengers connect at Chicago airport.

11. Although emissions per passenger-km decrease with the distance, absolute emissions increase with it.

12. 'Brussels Airlines and Jet Airways join forces with Thalys to link their networks to the centre of Paris, via Brussels Airport', 14 September 2011.

ACKNOWLEDGEMENT

This chapter is part of the DATE project, financed by the European Commission under a Marie Curie Intra-European Fellowship for Career Development. It reflects only the authors' view. The European Union is not liable for any use that may be made of the information contained therein.

REFERENCES

Behrens, C., & Pels, E. (2009). *Intermodal competition in the London–Paris passenger market: High-speed rail and air transport* (pp. 1–24). Amsterdam: Tinbergen Institute Discussion Paper TI.

Bonsall, P. (1996). Can induced traffic be measured by surveys? *Transportation*, *23*(1), 17–34.

Brueckner, J., & Spiller, P. (1994). Economies of traffic density in the deregulated airline industry. *Journal of Law and Economics*, *37*(2), 379–415.

Burghouwt, G., & de Wit, J. (2005). Temporal configurations of European airline networks. *Journal of Air Transport Management*, *11*(3), 185–198.

Campos, J., & de Rus, G. (2009). Some stylized facts about high-speed rail: A review of HSR experiences around the world. *Transport Policy*, *16*(1), 19–28.

Campos, J., & Gagnepain, P. (2009). Measuring the intermodal effects of high speed rail. In G. de Rus (Ed.), *Economic analysis of high speed rail in Europe* (pp. 71–88). Bilbao: Fundación BBVA.

Capon, P., Longo, G., & Santorini, F. (2003). Rail vs. air transport for medium range trips. Paper presented at the 7th International Conference on Traffic Science (ICTS), Nova Gorica, Slovenia.

Cascetta, E., Papola, A., Pagliara, F., & Marzano, V. (2011). Analysis of mobility impacts of the high speed Rome–Naples rail link using withinday dynamic mode service choice models. *Journal of Transport Geography*, *19*(4), 635–643.

Chang, J., & Lee, J.-H. (2008). Accessibility analysis of Korean high-speed rail: A case study of the Seoul metropolitan area. *Transport Reviews*, *28*(1), 87–103.

Cheng, Y.-H. (2010). High-speed rail in Taiwan: New experience and issues for future development. *Transport Policy*, *17*(2), 51–63.

Chester, M. V., & Horvath, A. (2010). Life-cycle assessment of high-speed rail: The case of California. *Environmental Research Letters*, *5*(1), 1–8.

Chi, A., & Crozet, Y. (2004). L'intermodalité TGV/Air: quel second souffle pour la gare de Lyon Saint-Exupéry TGV? *Transports*, *428*, 360–366.

Chiambaretto, P., Baudelaire, C., & Lavril, T. (2012). Measuring the willingness-to-pay of air-rail intermodal passengers. *Journal of Air Transport Management*, *26*, 50–54.

Chiambaretto, P., & Decker, C. (2012). Air–rail intermodal agreements: Balancing the competition and environmental effects. *Journal of Air Transport Management*, *23*, 36–40.

Coto-Millán, P., Inglada, V., & Rey, B. (2007). Effects of network economies in high-speed rail: The Spanish case. *The Annals of Regional Science*, *41*(4), 911–925.

de Rus, G., & Nombela, G. (2007). Is investment in high speed rail socially profitable? *Journal of Transport Economics and Policy*, *41*(1), 3–23.

Dobruszkes, F. (2001). Multimodalité TGV – avion: considérations sur le cas de Bruxelles. *Belgeo*, *200*(1/4), 335–350.

Dobruszkes, F. (2008). *Libéralisation et desserte des territoires. Le cas du transport aérien européen*. Brussels: Peter Lang.

Dobruszkes, F. (2011). High-speed rail and air transport competition in Western Europe: A supply-oriented perspective. *Transport Policy*, *18*(6), 870–879.

Dobruszkes F., Dehon, C., & Givoni, M. (2012). Does European high-speed rail affect the current level of air services? An EU-wide analysis. Paper submitted to the WCTR 2013 conference.

EC(European Commission). (1998). *COST 318: Interactions between high speed and air passenger transport (final report)*. Office for Official Publications of the European Communities, Luxembourg. Retrieved from http://temis.documentation.equipement.gouv.fr/documents/Temis/0030/Temis-0030323/12414_1.pdf

EC(European Commission). (2005). Community guidelines on financing of airports and start-up aid to airlines departing from regional airports. *Official Journal of the European Union, C 312*, 1–14. Retrieved from http://eur-lex.europa.eu

EC(European Commission). (2011). *White Paper on transport. Roadmap to a single European transport area – towards a competitive and resource efficient transport system*. Publications Office of the European Union, Luxembourg. Retrieved from http://ec.europa.eu/transport/themes/strategies/2011_white_paper_en.htm

Fourniau, J.-M. (1989). La génèse du TGV Sud-Est. Innovation et adaptation à la concurrence. *Culture technique*, *19*, 85–94.

Francis, G., Dennis, N., Ison, S., & Humphreys, I. (2007). The transferability of the low-cost model to long-haul airline operations. *Tourism Management*, *28*(2), 391–398.

Friebel, G., & Niffka, M. (2009). The functioning of inter-modal competition in the transportation market: Evidence from the entry of low-cost airlines in Germany. *Review of Network Economics*, *8*(2), 189–211.

Friederiszick, H., Gantumur, T., Jayaraman, R., Röller, L.-H., & Weinmann, J. (2009). *Railway alliances in EC long-distance passenger transport: A competitive assessment post-liberalization 2010*. ESMT White Paper No. WP-109-01. ESMT European School of Management and Technology, Berlin.

Givoni, M. (2007). Environmental benefits from mode substitution: Comparison of the environmental impact from aircraft and high-speed train operations. *International Journal of Sustainable Transportation, 1*(4), 209–230.

Givoni, M., & Banister, D. (2006). Airline and railway integration. *Transport Policy, 13*(5), 386–397.

Givoni, M., & Banister, D. (2010). *Integrated transport: From policy to practice.* London: Routledge.

Givoni, M., Dobruszkes, F., & Lugo, I. (2012). Uncovering the real potential for air-rail substitution – an exploratory analysis. In O. Inderwildi & D. King (Eds.), *Transport and the environment. Addressing the sustainable mobility paradigm* (pp. 495–512). London: Springer.

Givoni, M., & Dobruszkes, F. (2013). A review of ex-post evidence for mode substitution and induced demand following the introduction of High-Speed Rail. *Transport Reviews* (forthcoming).

Harvey, D. (1989). From managerialism to entrepreneurialism: The transformation in urban governance in late capitalism. *Geografiska Annaler, Series B, Human Geography, 71*(1), 3–17.

Jiménez, J. L., & Betancor, O. (2012). When trains go faster than planes: The strategic reaction of airlines in Spain. *Transport Policy, 23*, 34–41.

Klein, O. (2001). La genèse du TGV, une innovation contemporaine de l'épuisement du fordisme. *Innovations, Cahier d'économie de l'innovation, 13*(1), 111–132.

Lee, J.-K., Yoo, K. E., & Jung, S.-Y. (2012). A study on the effect of high-speed railway launch to the air passengers' mode choice behaviour. Paper presented at the 16th ATRS World Conference, Tainan, Taiwan, June 2012.

Miyoshi, C., & Givoni, M. (2013). The environmental case for high speed train in the UK: Examining the London–Manchester route. *International Journal of Sustainable Transportation, 8*(2), 107–126.

Mokhtarian, P., Samaniego, F., Shumway, R., & Willits, N. (2002). Revisiting the notion of induced traffic through a matched pairs study. *Transportation, 29*(2), 193–220.

Powell, E. (1997). The TGV project: A case of technico-economic dirigisme? *Modern & Contemporary France, 5*(2), 199–214.

Rothengatter, W. (2011). Competition between low-cost airlines and high-speed rail. In R. Macário & E. Van de Voorde (Eds.), *Critical issues in air transport economics and business* (pp. 319–342). London: Routledge.

Steer Davies Gleave. (2006). *Air and rail competition and complementarity.* Final report for the European Commission (DG TREN). Steer Davies Gleave, London.

Vespermann, J., & Wald, A. (2011). Much ado about nothing? – an analysis of economic impacts and ecologic effects of the EU-emission trading scheme in the aviation industry. *Transportation Research Part A: Policy and Practice, 45*(10), 1066–1076.

Vickerman, R. (1997). High-speed rail in Europe: Experience and issues for future development. *The Annals of Regional Science, 31*(1), 21–38.

Wardman, M., Bristow, A., Toner, J., & Tweddle, G. (2002). *Review of research relevant to rail competition for short haul air routes.* Report commissioned by Eurocontrol. Eurocontrol Experimental Centre, Bretigny Sur Orge. Retrieved from http://www.eurocontrol.int/eec/public/standard_page/DOC_Report_2002_011.html

CHAPTER 9

AERIAL EMERGENCE: CRISIS MANAGEMENT AND THE SUSTAINABILITY OF EUROPEAN AIRSPACE

Peter Adey

ABSTRACT

Purpose – *The chapter explores the sustainability of European aeromobility and the socio-technical systems that enable it by examining how certain threats, risk and uncertainties are managed in European airspace.*

Methodology/approach – *The chapter examines the discussions around the inception and deployment of methods of 'crisis management' to airspace governance and focuses on the importation of particular systems of knowledge, practices and technologies from other contexts.*

Findings – *The chapter finds that European authorities have installed a combination of capacities involving central and decentralised coordination and decision making – as well as training and simulation – in order to better anticipate and respond to moments of crisis, although the*

Sustainable Aviation Futures
Transport and Sustainability, Volume 4, 199–215

ISSN: 2044-9941/doi:10.1108/S2044-9941(2013)0000004009

politics of such an arrangement is not unproblematic to the interests of state sovereignty and the airlines' suspicion of regulation or lack thereof.

Originality/value – *The chapter is one of the first to make explicit how ash-cloud crises and disruption to airspace are being governed post the more immediate consequences of the 2010 ash cloud.*

Keywords: Emergency; crisis management; Europe; airspace; contingency planning

INTRODUCTION

It is one year on from the 2010 ash-cloud crisis which saw much of Europe's airspace disrupted for over a week and 10 million passengers stranded throughout the world. The same thing is happening again. Except this time it is not for real. This is exercise Volcex 1.0. The disruption is held on the displays of the screens of Europe's airspace managers at EUROCONTROL (see Fig. 1). Their perspective is augmented by new and improved visualisation tools to gain a firmer grip on the ash-cloud's movements and, importantly, on the coordination of the institutional and organisational complexities and resources required to respond to the crisis. The exercise

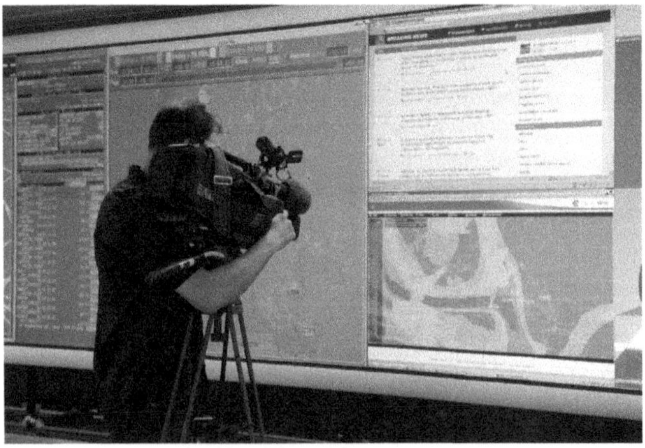

Fig. 1. The Media Report from the Volcex Exercise.

has been designed to test a new structure of emergency or crisis management of European airspace following the *crisis* the ash-cloud event brought to the air-travel and airspace industry as both intense disruption and catalyst for organisational change. Curiously in some of the press photos a map of the no fly zone above Libya is displayed, an image we will return to later.

In the world of aviation we are used to talking about a crisis of events, as security efforts performed under various underlying logics work in the context of international terrorism and other kinds of threats, such as disease and health insecurities (Ali & Keil, 2006; Warren, Bell, & Budd, 2010). Of course those studies have tended to explore the development of airport and airline security practices with particular emphasis on the corporeal experience of the airport passage – from full body x-ray scanners and invasive scrutiny to the pernicious intrusions on people's privacy and their life chances (Amoore & Hall, 2009). Such practices have been shown to be intensely networked by forms of surveillance and information sharing of data which support those activities. At the same time, the regular deployment of various meteorological metaphors to describe the instability of the challenges brought by intensified security, regulation or contemporary economic pressures emphasise a sense that aviation is moving through stormy weather that is not necessarily in its own hands or of its own making (Goetz, 2004; Leinbach & Bowen, 2004). What I am trying to get at is the sense that we are less used to discussing a 'crisis' or 'emergency' of aviation that is internal to the system. In other words, in the context of the networked disruption that can threaten the sustainability of aviation, what happens when the crisis is born from the network of air-transport itself or, more precisely, from the vulnerability of a complex and interdependent system to handle particular sorts of challenges like an ash-cloud crisis?

This chapter explores head-on the sustainability of aeromobility through questions around how the nature of threat, risk and uncertainties are managed in European air-travel. Such a focus means turning to the sustainability of the socio-technical system that supports that mobility. In particular it directs us to an understanding of a disrupting, unruly, physical and social infrastructural setting like an ash cloud encountered by the aviation system. In the context of the air-travel and other events, the system is actually peculiarly resilient. Despite the appearances of various shocks and crisis in international aviation, the system is notably resistant to disruptive threats perhaps because it is so 'systemic'. If part of the system fails, or damage occurs, the system is flexible enough to work around it. It has to do this, for instance, when holes in the airspace system appear because a country's air-traffic controllers go on strike. However, when disruption has escalated beyond the capacity of one state or organisation to manage it, as in the 2010 crisis, rather than proving resilient the system found itself subjected to (almost) cascading failures (Little, 2002) in the stubborn and inefficient ability to coordinate itself properly (Adey & Anderson, 2011). Moreover, on these occasions when disruption is more wide-scale, societies have been noticeably reluctant, not very good at, or really quite unused to getting around and organising themselves differently (Birtchnell & Büscher, 2011).

The way I want to expand on these problems is through the way that the issues revealed during the 2010 events have led to a series of changes to the coordination of European airspace and even wider systems of mobility. While the crisis has been relatively well discussed so far in the literature in numerous articles and special issues, I draw attention to how the crisis and the later changes in airspace management have been understood through quite a particular and yet familiar set of approaches in the wider way that liberal societies now handle or govern risks. That is, within a logic of handling the onset of an emergency or crisis through intense coordination and preparation, the logic is one of preparedness and the practice of 'crisis' or 'emergency management' (Anderson, 2010).

AERIAL LIFE AND 'EMERGENCY'

Part of where we could begin this chapter is in relation to a set of arguments I have made elsewhere about the nature of aeromobility and our aerial lives (Adey, 2010). There we might suggest how the increasingly central role of air-transport mobilities to our society has led to the genesis of what we could call an 'aereality' – a distinctive kind of mobile society, a systemic 'life on the move' – which the aeroplane has worked to produce (Cwerner, Kesselring, & Urry, 2009; see also Sheller & Urry, 2000 on the car). The rub is, the reliance of life upon these *systems* of movement makes that life particularly vulnerable to blocks, breaks and congestion within the system. Path-dependency as Urry shows in the context of automobility is quite difficult to shift. Thus it is the very systems of mobility, sociability and connectivity; characterised by contingency, circulation and complexity as Mick Dillon might argue, that we build through mobility by air. This dependency renders the vulnerability of that life open to the disruptions of that very system, and critically *through* that very system. What this revealed in the ash-cloud crisis for passengers, was that there were very few ways to get home beyond air-travel. Or at least the alternative networks to aeromobility were quickly booked up and couldn't cope with the demand. What's more, there was little coordinated effort to be able to offer alternative forms of mobility were they needed. All this meant that people were 'stranded' – some for one and even two weeks. For many, this situation had been inconceivable and its occurrence placed severe pressures on the kinds of relations that compose our lives which require mobility for proximity, from keeping appointments to maintaining family relationships, child care, shopping, going to the doctors and getting to work.

Perhaps we had become so used to air-travel's minor inconveniences and disruptions that sustained disruption had not been thinkable. Indeed, it would not be within the realms of our normal habits and routines to prepare for such eventualities either. To travel with the likelihood that one could be stranded for one or two weeks would probably mean ensuring different kinds of contingency arrangements should one get stuck, despite the fact that air-travel has not always been so consistent or reliable. Flight's early years saw endless descriptions of variable delays, unforeseen layovers which were then considered mere nuisances in a more glamorous and pioneering age of aviation (Wohl, 1994). The original establishment of long-haul flights along colonial air-routes could suffer days and sometimes many weeks of disruption (Pirie, 2009). Where does this take us? Well first, our mobile lives, becoming premised upon a broadly reliable air-transport system, finding themselves unable to return, were heavily disrupted. And second, and where this chapter will explore in more depth which led to the first, the system of air-travel itself seemed unable to cope with a particularly intensive and extensive complex event.

I am not trying to suggest that such an interaction of the kind of complex networks that constitute the air-travel system creates a system instability or vulnerability always prone to fail. Because even when it goes wrong we see examples of considerable system resilience in routine network congestion and challenges. In other cases we see the remarkable coming together and performance of different actors in failure and success in the 'miracle on the Hudson' which has not been simply revealed as a miraculous occurrence located within the purposeful actions of the aircraft's Captain Sullenberger, but the working relationship of a complex series of diverse actors, technologies and software.

Within the world of air-travel there are competing ideas on the argument of where risk is to be located. Following 9/11 many perspectives turned to the seemingly ungoverned or rather complex organisational arrangement of air-transport's security practices that have followed the de-regulation and liberalisation of the air-transport sector to the private sector. In this sense the organisational complexity of the provision of airport and airline security was seen to have decreased at least the financial burden of aviation by distributing it across the system. Instead, the difficulty of instituting consistent standards of practices, staff requirements and training, protocols, along with an array of other issues, may have led the system to be made more vulnerable to threat.

By apparently moving away from 'security' to the management of airspace, my analysis is explicitly informed by work which has argued that the paradigmatic figure of threat *is* the networked disruption and that an array of anticipatory techniques and logics have been invented in order to stop such disruptions before, as they, and once they have occurred. Terrorism, trans-species epidemics and other threats are not separate from normal life in liberal democratic societies; they are not seemingly exogenous challenges distant on the horizon, but they might incubate within, are ready to erupt or require little to set them going (Cooper, 2010). Their conditions of formation are the very conditions that support highly mobile ways of living.

What this all means for the management of aerial crises is a convergence of the techniques we see employed in other areas of societal risk: that is crisis or emergency management processes which employ a logic not to necessarily mitigate or prevent the advent of an event, but rather to act following an occurrence and seek to manage how their consequences unfold. Within the context of the United Kingdom an apparatus of civil protection known as 'civil contingencies' and, in the United States 'distributed preparedness', has obliged many statutory duties. The legislation operationalises a relatively precise understanding of an 'emergency'[1] set into action by the advent of a precipitous event (Anderson & Adey, 2012). UK Civil contingencies prescribes an array of risk assessment, emergency planning, training and exercising in order to meet a series of defined risks, hazards and threats that come to, importantly, threaten life. These measures do not seek necessarily to stop a flood, bomb or chemical spill from occurring but act to halt the emergency in its flow, to nip it in the bud as if a firefighter would stop a fire from spreading. Although the logic is rather convoluted and coexists with others such as precautionary or preemptive logics, it privileges a relation towards the future that explicitly emphasises response and the coordination of that response. Moreover, it articulates a multi-sited and multi-scalar organisational or agency structure with particular responsibilities to prepare for given events.

Let us turn now to the ash cloud in order to work through the nature of the emergency that has loomed large in the management of European airspace, and the following institutionalisation of crisis governance within its organisation.

ASH CLOUDS, CRISIS AND SYSTEMIC DISRUPTION

Exercise Volcex simulated a continuous eruption of the Icelandic volcano Grimsvötn starting at 0001UTC on 12 April 2011 and was led by Nigel Gait of the London Volcanic Ash Advisory Centre and EUROCONTROL. The exercise had several functions, particularly to test the effectiveness of the newly amended ICAO volcanic ash contingency plan, and even more notably to assess the efficacy of a newly created administrative unit in European airspace: the European Aviation Crisis Coordination Cell (EACCC). The EACCC would emerge as a potential solution to the events in 2010 amid the criticism levelled at EUROCONTROL and individual states. As the problem was posed at the time, there had not seemed to be enough coordination of the response to the event. Since Volcex there have been a subsequent number of exercises coordinated by EUROCONTROL and the ICAO (Volcex Steering Group, 2012), but before we get into the detailed function of the EACCC and the workings that the exercise sought to perform and test, it is worth dwelling on what the revisions to Europe's airspace management sought to achieve and why.

The Eyjafjallajökull ash-cloud disruption essentially revealed a particular kind of vulnerability. That is to say, even while the aviation system is pretty resilient to different kinds of disruption, it demonstrated what Budd, Griggs, Howarth, and Ison (2011) and others have begun to show was a series of issues that displace our sense of 'crisis' away from the volcanic eruption as natural disaster to somewhere else (Donovan & Oppenheimer, 2010). Rooting it within the public politics and policy that constructed the debate, this, they argue, was about the 'failed management of European airspace which might be addressed by better planning and human agency' (Budd et al., 2011). Three broad issues were identified that the following changes sought to address. First, was speculation.

To all intense purposes the type of disruption that took place had not been anticipated by the aviation industry or passengers. That is not to say that ash-cloud events and their consequences are not routinely felt because of course they are in Indonesia, Alaska, the Philippines, Chile, New Zealand and elsewhere, or that the specific volcanic event had not been anticipated within scenarios and contingency plans. The issue was the extent and scale to which the event was able to disrupt the system in a way that was difficult to predict because of its complexity and the limited powers of governance any organisations held over the response. There simply was not the authority to anticipate the reaction of individual states to those events.

Second, was coordination. Much of the criticism has been leveled at what was seen to be the poor leadership and coordination of the crisis. The response of states seemed to be patchy. Some countries immediately declared no fly zones whilst others did not. Many used zero-tolerance limits while airlines called for those assumptions to be revised. Airlines and the IATA complained at the lack of speed in the decision making. In all, the lack of direction, leadership and effective collaboration were heavily criticised. As Brian Flynn, Network Manager at EUROCONTROL explains it:

> if the airspace of one country is closed and the airspace of another is open, yet the same phenomena exist in both States, that's an illogicality in the network. So we need to have clear, uniform procedures and decision making. (Butterworth-Hayes, 2012)

Third, knowledge. Take the manner in which the cloud's presence was first felt through a set of affective and not necessarily calculative registers. Through a series of uncertainties, feelings rose to the foreground not as consequences, as Donovan and Oppenheimer rightly describe but perhaps incorrectly locate, but rather as incipient systemic disruptions. The force came from the cloud's 'felt potential'. Even though no aircraft had actually crashed, speculation over the ash and its dangerous *potential* effects upon aircraft overshadowed many reactions. The future consequences of a crash were present in the here and now as worries, doubts and anxieties about air-travel were threatened by a natural phenomenon that appeared indifferent to human life. The ash-cloud crisis revealed a severe problem in actually knowing what was happening to the ash-cloud itself, where it was, where it was moving, what densities or concentrations of ash-particulates it had suspended and what possible effect they could have on aircraft. All the more severe was the difficulty in relaying that information not only to the public but to the network and airspace managers and decision makers who needed accurate and up-to-date information. It was particularly difficult to get a handle on an amorphous and indeterminate shape, and especially difficult to visualise it effectively in relation to airspace and the complex aircraft routes which have posed their own difficulties to representation.

While each of these themes could be explored independently of one another, the rest of the paper folds the discussion through the changes to European airspace management since 2010 which have been primarily concerned with crisis coordination.

COORDINATION

Almost immediately following the 2010 ash-cloud crisis Europe's airspace managers announced that a European Airspace Crisis Coordination Cell would be setup to plan and act in advance of the event of a crisis. Setup within EUROCONTROL and with EUROCONTROL really at the helm as Network Manager given its new position granted by the European Commission, the EACCC has several functions but its primary role is 'to support coordination of the response to network crisis situations impacting adversely on aviation, in close cooperation with corresponding structures in States'. In other words, the rationale for the cell picks up on several main issues. The initial response was deemed to have been uncoordinated because it lacked systematic cooperation between States who had appeared to take matters into their own hands in the management and subsequent closure of their airspace. As opposed to the 2010 eventualities, the EACCC should be activated and proceed in the following sequential and proportionate escalation of resource use, decision making and cooperation. First, once a 'crisis' has been named the EACCC should communicate with 'State level focal points' and those at risk. It should develop a crisis-mitigation policy in order to try to contain or stop the crisis in its tracks, notably by liaising with network structures at a State level. This includes the 41 national civil aviation authorities under the area of operations of EUROCONTROL. Once a crisis is resolved, the EACCC will stand-down and it will debrief key partners to determine lessons to be learned and fold those lessons back into the planning cycle.

Of course the important aspect of a coordination cell is that it is not always in operation. Rather crisis or emergency conditions should escalate to the point that the cell is activated. This means that the organisation of European and Northern Atlantic airspace is now managed in a way that draws a distinction between normality, 'normal disruptions' and those of 'crisis'. Network disruptions are defined as those that 'impact the network operations which may be escalated to a crisis', thus disruptions are really conditions out of the ordinary which have the potential to become crisis-like, or crisis is imminent. Alternatively, the escalation to a crisis is defined as:

> a state of inability to provide air navigation service at required level resulting in a major loss of network capacity, or a major imbalance between network capacity and demand, or a major failure in the information flow in one or several parts of the network following an unusual and unforeseen situation.

What is crucial to this definition is how disruption and crisis are viewed at a network level. Neither does it matter what the nature of the precipitating cause actually is. The contingency plans consider a list of possible events that could lead to particular sets of networked problems. These are: bomb threats; terrorist threats, nuclear contamination, possible war; chemical contamination, fire, airborne diseases, pandemics; possible pan-European function failure; industrial action of several Air Navigation Service Providers (ANSPs) and cyberattack. In fact, cyberattack is one of the biggest and unresolved concerns for both EUROCONTROL and national security strategies.

What matters to this framework is the potential effect of events upon the network. Crisis conditions are achieved because of the potential damage upon the network and, furthermore, unpredictable and unforeseen circumstances should information flow be disturbed. The implication of this structure is the sense that crisis can emerge from disruptions that may well happen in one State or area of the network but could achieve significant impacts in other parts of Europe and beyond that. The sense that the crisis escalates beyond territorial boundaries and the management processes of a network has meant that several other related criteria should be considered before activating the cell. These include an event in one or more States/FABs (functional airspace blocks) that causes significant flight cancellations and displaced/stranded passengers, or may severely affect air traffic management (ATM) performance. There are issues of longevity such as when the event is predicted to last for more than 24 hours, or is likely to recur. There are issues of unpredictability, when the duration of the event cannot be predicted with much certainty, or when the response is non-standard, that is between civil/military operations across a State border. Finally, a cell may be activated when the event may require far closer coordination of ATM and other transport modes.

During Volcex 2011, a frequent example of disruption to airspace management was the UN legislated no-fly zone above Libya which required non-standard response in order to reroute aircraft around the airspace from North Africa and Southern Europe. The EACCC was not actually activated, but the Network Manager had to assess whether it should be given the potential disruption to over-flights whose flight paths were diverted around the Libyan Flight Information Region (FIR), via Tunisia and Algeria to the West and Malta, Greece and Egypt to the East. What complicated this was that the adjusted routes became affected by the industrial action of air traffic personnel in Greece, which further hindered flight traffic from the gulf that went via the Mediterranean – in-effect a result of the networked dependency of a few important air routes (Lian, 2010). Adding fragility to the networked vulnerabilities of European airspace, both Libya and the reduction of capacity in Greece illustrated the potential for the network to be affected from multiple points that then spread disruption to other parts. Rerouting was not only costly adding hundreds of miles to flight routes and severe delays to the East via Turkey, but the displaced traffic also threatened to create congestion in the areas they were diverted to.

VISUALISATION

The ash-cloud's presentation as an eventual disruption by both the media and public commentators was anomalous, uncertain, mobile, threatening. Indeed, the turbulence of the cloud resembles the kinds of scenarios that domestic civil contingencies planners use as they design their scenarios for multi-agency exercises. This is not because turbulent clouds are highly regular occurrences, although the UK Buncefield oil refinery disaster is one example, as was a tornado in Birmingham in 2005, but because the characteristics of their movement simulates the sorts of cascading effects which other sorts of events will produce. What's more, because of this character, they challenge responders by unpredictably moving across agency jurisdictions, between and through natural and man-made capillary like systems. As they move, they escalate the nature of their consequences and the necessary response required to govern them. Thus it is not hard to see how the Volcex exercises provide useful events to prepare for multiple sorts of eventualities that could potentially disrupt the network. And yet, the cloud's intangibility – Lily Ford (2010) recognises the irony that from the ground the cloud was invisible just as its name was unpronounceable – caused other kinds of problems.

Within European aviation crisis management one of the key innovations has been a method of visualising the presence, shape, location and concentration of a potentially disruptive ash cloud. This has materialised in the European Crisis Visualisation Interactive Tool for Air Traffic Flow and Capacity management (EVITA) (see Fig. 2). EVITA does several things but in the European Civil Aviation Conference Region it primarily integrates data from the regional Volcanic Ash Advisory Centres in London and Toulouse. One of the distinctive problems of the ash-cloud crisis was that the data from London describing the cloud's spread, movement and density from Iceland was not easily integrated into other Advisory centre

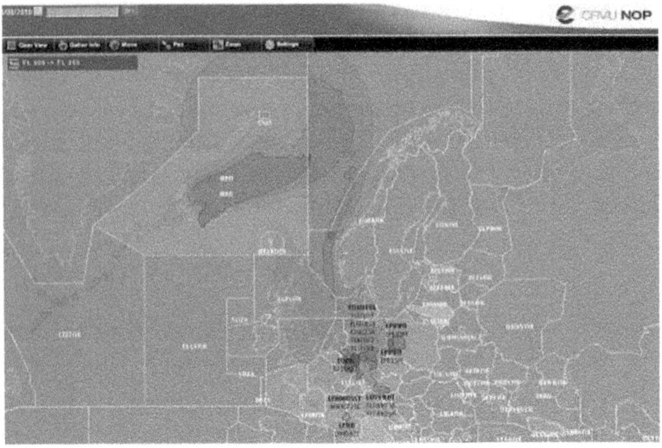

Fig. 2. The EVITA Visualization Tool.

data, nor was it brought into easy relation with European airspace management maps so that airline operators and network managers could easily plot individual flight routes in relation to the VAAC data. This technological capacity to visualise the total extent of an ash cloud in relation to airspace and its utilisation, can be compared to other sorts of sensibilities to an ash-cloud encounter discussed later.

RESPONSIBILITY

The new arrangements importantly assign a variety of stakeholders with a series of responsibilities and remits of action. At the head of the EACCC is the Network Manager who will chair the EACCC's response to the crisis. The EACCC is then tasked with: coordinating the response; supporting activation of contingency plans at a member state level and mitigating measures to ensure that the crisis cell can respond in a timely fashion. This includes ensuring that the sharing of information is fast and complete to support decision making and risk management. One of the key complaints during the 2010 crisis was the apparent slowness with which states were able to make decisions, following what appeared to be an initial snap-decision to close their airspaces. Within the new EACCC arrangements, there are now procedures to be able to begin a teleconference with all of

Europe's airspace managers in the space of an hour. Speed is of the essence to support timely but considered cooperation and coordination.

Under these arrangements the EACCC should also work with the European Commission to liaise with other transport modes to provide alternative modes of mobility or transport. This was a key problem during the 2010 crisis when European railway networks, among other modes of travel, were unable to provide the sufficient capacity. It was even mooted that naval ships would be sent to the United States, but were used to transport stranded travellers in Southern Europe instead.

Furthermore, in terms of the composition of the cell, the cell represents a higher-level multi-organisational coordination through which decisions can be both cascaded outwards to their representative agencies and others. The EACCC thus includes a representative from the Member State holding Presidency of the Council, the European Commission, the EASA, EUROCONTROL, the military, air navigation service providers, airports and airspace users. This means that the structure integrates with more local contingency planning in order to distribute information and decisions to and from those structures.

DISTRIBUTION

As with the coordination of most civil emergencies, emergency management determines a series of rationalised sequences of action in the advent that action is needed. The EACCC is no different to this and has developed a series of procedures for decision making at multiple levels, where we might consider the system as affording both centralised *and* distributed coordination.

In relation to these procedural moves (EACCC, 2012), the EACCC draws upon a familiar understanding of the timeline of a crisis or emergency and associated intensities and scales of action relevant to each phase. One of the most important phases is the pre-alert phase, which despite the logic of preparedness 'crisis management' lives by, simultaneously exists alongside other forms of mitigation and preemption. The pre-alert phase within EACCC means that important actions can be made to nip a crisis in the bud, or plans can be put in place to anticipate potential eventualities, should they happen. Since the EACCC was launched, a notable case to concern the cell was Japan's Fukushima nuclear meltdown in April 2011, which sent the cell into pre-alert given the potential disruption of the radioactive cloud, were it to move in significant quantities to affect European Airspace (Butterworth Hayes, 2012). In that instance, the cell made contact with nuclear scientists in Europe to anticipate its potential consequences on European aviation and determine a series of responses. The EACCC is also retrospective, in that will fold any lessons learnt from a particular event back into its cycle of preparedness. Thus, following an event and a debriefing of analysis of the cell's functions, a lessons learned report will be filed. The Network Manager and the EACCC will develop actions and monitor the progress of those actions for the future improvement of the process. From pre-alert, the management of the crisis is assumed to move through a disruption phase, to crisis to recovery and analysis with each phase associated with varying degrees of intensity of activity and cooperation.

Like most civil emergency management techniques, coordinated cooperation works across multiple scales with subsidiarity given much higher priority than it might at first seem. In the new arrangements for volcanic ash, new ICAO (2012) guidelines pair the EACCC's role with the primacy of national civil aviation authorities and airline operators to make a risk-based and local decision according to established protocols of hazard identification, risk severity, risk probability, tolerability, mitigating actions and the keeping of documentation and records (see also CAA, 2011). The lowest level of this decision making over an individual flight is given to the actual aircraft themselves. The new recommendations rehearse a series of indications that could be expected by a flight crew during an actual encounter that may be preceded by the visual sighting of a cloud. Those are 'odour', in the form of a smoky or acrid smell; haze, that will actually develop in the flight cabin and settling minute particles of dust; changing engine conditions such as 'torching from the tailpipe, and flameouts'; airspeed, changes in cabin pressurisation and static discharges.

While the ultimate validation of the new plans can only occur when faced with a real 'crisis' event, these emergency coordination responses and procedures are regularly exercised. Indeed, the ICAO's central recommendations centre around effective coordination 'between all concerned' and the annual volcanic ash exercise or simulations that states and their operators are encouraged to attend. Maintaining a level of preparedness through repetitive training is the EACCC's fourth area of responsibility in non-crisis periods.

Volcex 12/01 was designed to exercise the scenario of a volcanic ash plume from Sao Miguel Island in the Azores that would impact flights in the Mediterranean, routes from the Adriatic to Eastern Europe, to Africa and long-haul journeys from North America to the Middle East and India

(Volcex Steering Group, 2011). The scenario would be designed to effectively test the use of special air-reports on volcanic ash, contingency plans – especially the activation of the EACCC – and the enrolment of other countries such as Morocco, Algeria and Tunisia.

CONCLUSIONS

What can we conclude from this discussion in relation to the sustainability of air-travel and the current system of organising airspace which depends on so many actors, institutions and technologies?

The event of the ash-cloud disruption and the following institutional changes should gather together more focused debate about how the nature of risk is understood and acted upon by European airspace decision makers, particularly troubling any distinction between an initial event and its secondary consequences. This is effectively about understanding the complex interdependencies that circulate networks inside and outside the aviation world. It is precisely this understanding that is demonstrated in Europe's new arrangements which have instituted a crisis management approach to airspaces borrowing practices from many private, civil and state organisational techniques employed to manage or govern crises and emergencies – more broadly known as resilience. The way this works is to bring multiple organisations and networks together in order to anticipate and respond to particular events.

While the appearance of Europe's airspace under this system appears to be one of centralised control it is clearly far more nuanced than that and it would be a mistake to level familiar accusations of a centralised and hierarchical system of governance at the European authority. Coordination of the system actually appears to entail structures which responsibilise and distribute that responsibility both to predefined protocols and plans held at a variety of scales, from airline operators, to individual nation-states, to the Commission. There is also a degree of responsibility to improvise and discuss at speed, with far more up-to-date information and knowledge about the event than was before possible. Furthermore, Europe's crisis management delegates far more of the decision making to the airlines themselves and the states that regulate their assessment of the risk. In short, while coordination is held within a centralised albeit temporary body, autonomy for decisions is simultaneously distributed to almost the lowest level possible, the aircraft.

If the 2010 ash-cloud disruptions and crises pointed towards a potentially more fragile aviation future than we might have imagined, a future of uncertain and unstable airline bookings, questions over the effectiveness of our insurance coverage, the ability of the airlines to care for their customers, and for passengers forced to navigate far more complex schedules, logistics and decision making than they are used to as multiple systems broke down, then perhaps the sustainability of aeromobility should be brought into more sustained questioning, together with ongoing threats to the aviation industry from rising fuel prices and oil scarcity. Furthermore, with a potentially more uncertain climate and more variable weather conditions including more extreme variations in temperature and rainfall in Europe, not to mention the possibility of these effects on other volcanic eruptions, or even recurring civil and geopolitical unrest affecting airspace use and flight paths, just how sustainable is air-travel? Interestingly, what the developments in European airspace crisis management approaches show is a more supple system able to respond or adapt to those changing conditions. The propensity for the system is not to snap or reach breakdown, but the promise is perhaps one of a coordinated, yet relatively decentralised management of airspace and airlines. The premise of such an approach is not to remain blind to these aviation futures, but to try to foreclose their potential disruption by imagining them, anticipating them and acting upon them as and when they happen. Such a system seems able to shift and adjust to these events, beyond and around the literal and metaphorical incursion of an uncertain cloud of disruption creeping its way towards our skies.

NOTE

1. The apparatus has gone through several stages of review motivated by the onset of challenging events. This was not merely from 9/11 but major incidents such as Foot and Mouth, the Fuel Protests of 2000 and an evolution of policy since the emergence of the Civil Contingencies Unit called into being in the 1970s in response to the dock and coal workers strikes that necessitated several 'states of emergency' in order to govern threats to essential services.

REFERENCES

Adey, P. (2010). *Aerial life: Spaces, mobilities, affects.* London: Wiley.

Adey, P., & Anderson, B. (2011). Anticipation, materiality, event: The Icelandic ash cloud disruption and the security of mobility. *Mobilities*, *6*(1), 11–20.

Ali, S. H., & Keil, R. (2006). Global cities and the spread of infectious disease: The case of severe acute respiratory syndrome (SARS) in Toronto, Canada. *Urban Studies*, *43*(3), 491–509.

Amoore, L., & Hall, A. (2009). Taking people apart: Digitised dissection and the body at the border. *Environment and Planning D: Society and Space*, *27*, 444–464.

Anderson, B. (2010). Preemption, precaution, preparedness: Anticipatory action and future geographies. *Progress in Human Geography*, *34*(6), 777–798.

Anderson, B., & Adey, P. (2012). Governing events and life: 'Emergency' in UK civil contingencies. *Political Geography*, *31*(1), 24–33.

Birtchnell, T., & Büscher, M. (2011). Stranded: An eruption of disruption. *Mobilities*, *6*(1), 1–9.

Budd, L., Griggs, S., Howarth, D., & Ison, S. (2011). A fiasco of volcanic proportions? Eyjafjallajökull and the closure of European airspace. *Mobilities*, *6*(1), 31–40.

Butterworth-Hayes, P. (2012). At the heart of crisis management. *Skyway*, (spring), 36–40.

CAA. (2011). Guidance regarding flight operations in the vicinity of volcanic ash. UK Civil Aviation Authority.

Cooper, M. (2010). Turbulent worlds financial markets and environmental crisis. *Theory, Culture & Society*, *27*(2–3), 167–190.

Cwerner, S., Kesselring, S., & Urry, J. (Eds.). (2008). *Aeromobilities*. London: Routledge.

Donovan, A. R., & Oppenheimer, C. (2010). The 2010 Eyjafjallajökull eruption and the reconstruction of geography. *The Geographical Journal*, *177*(1), 4–11.

EACCC CDM. (2012). European Aviation Crisis Coordination Cell (EACCC) Rules of Procedure. Final version 4.1, EUROCONTROL.

Ford, L. (2010). Mapping the clouds. *Fortean Times*, May. Retrieved from http://www.forteantimes.com/features/commentary/3475/mapping_the_clouds.html

ICAO. (2012). Flight safety and volcanic ash: Risk management of flight operations with known or forecast volcanic ash contamination, first edition, DOC 9974, ANB/487, International Civil Aviation Organisation.

Leinbach, T., & Bowen Jr., J. (2004). Airspaces: Air transport, technology, and society. In S. Brunn, S. Cutter, & J. W. Harrington Jr. (Eds.), *Geography and technology* (pp. 285–313). Houten, AK: Springer.

Lian, J. I. (2010). Network dependency and airline competition Consequences for remote areas in Norway. *Journal of Air Transport Management*, *16*(3), 137–143.

Little, R. G. (2002). Controlling cascading failure: Understanding the vulnerabilities of interconnected infrastructures. *Journal of Urban Technology*, *9*(1), 109–123.

Pirie, G. (2009). *Air empire: British imperial civil aviation, 1919–39*. Manchester: Manchester University Press.

Volcex Steering Group. (2011). Summary of the Planning Meeting of the First European and North Atlantic Volcanic Ash Exercise in 2012, EUR/NAT VOLCEX12/01, Lisbon, 6 December.

Volcex Steering Group. (2012). Summary of Discussions of the European and North Atlantic Volcanic Ash Exercises Steering Group, EUR/NAT VOLCEX/SG, Ninth meeting, Toulouse, 19 November.

Warren, A., Bell, M., & Budd, L. (2010). Airports, localities and disease: Representations of global travel during the H1N1 pandemic. *Health & Place*, *16*(4), 727–735.

Wohl, R. (1994). *A passion for wings: Flight and the Western Imagination*. New Haven, CT: Yale University Press.

PART III
PROSPECTS FOR CHANGE

CHAPTER 10

COALITION, AVIATION AND THE DESCENT TO 'POLITICS AS USUAL'

James Connelly

ABSTRACT

Purpose – *To examine the general political structure and background to decision making in aviation policy.*

Design/methodology/approach – *Largely conceptual and theoretical, applied to political and historical examples.*

Findings – *That 'politics as usual' signifies a structure characterised by adherence to underlying presuppositions which guide political and economic action and that moves towards different ways of thinking and acting and that environmental policy faces an uphill battle in the light of these often unacknowledged and typically unquestioned presuppositions.*

Practical implications – *Political action requires an accurate understanding of the circumstances in which it is to be undertaken. This chapter provides a clue to that proper understanding and hence is valuable for the appropriate orientation of political action.*

Sustainable Aviation Futures
Transport and Sustainability, Volume 4, 219–238

ISSN: 2044-9941/doi:10.1108/S2044-9941(2013)0000004010

Originality/value – *The chapter employs an everyday term, 'politics as usual', and analyses its structure, claiming that it identifies a real underlying set of default positions in politics.*

Keywords: Political theory; environmental politics; aviation policy

INTRODUCTION

Contemporary politics seems to be locked in a permanent oscillation between hope and fear: hope that an incoming government will live up to its claims and fear that it will not. This is matched by an oscillation between apathy and antipathy: as enthusiasm wanes, some become indifferent and others hostile. Despite appearances, political apathy and antipathy to politics and politicians are much the same. Whether one considers the hopes inspired by the election of President Obama or the stated aspirations of the United Kingdom's Conservative–Liberal Democrat coalition created in June 2010, one sees much the same pattern. The broader issue is whether this is merely the psychologically inevitable consequence of over-inflated hopes or whether there are deeper political structures and processes to consider. What, then, is it reasonable to expect of a new government or leader? Is there any way in which electorate, government or leaders can manage their expectations to ensure that idealism doesn't sour into cynicism? What are the limits to political change and can incoming governments ever step faster than a politically normal incrementalism dictates? These are large questions and this chapter cannot hope to answer them all; however, through an examination of the United Kingdom's Conservative-Liberal Democrat government, it will suggest that analysis reveals a typical underlying structure constraining the political opportunities and possibilities available, independently of the will or rhetoric of leaders.

For a brief and heady time it seemed plausible that the Conservative–Liberal Democrat coalition, formed in 2010, was, or could be, what it professed to be, namely 'the greenest government ever'. However, there were always compelling reasons for scepticism (Connelly, 2009, 2011). In this chapter I shall employ the term 'politics as usual' to describe the dominant mode of political life. This might sound cynical, but what follows is sceptical, not cynical. The difference I take to lie in the fact that cynicism and optimism are both forms of a priori judgement, differing only in the colour of the spectacles worn by the observer. Neither is genuinely based on evidence; rather, they bring their conclusions to the evidence and interpret it accordingly. A cynic is professionally disillusioned and proud of it. Scepticism, by contrast, rests on the need to evaluate evidence and to consider events, facts and decisions in relation to the underlying structure of political thought and action.

The chapter begins with an account of the constitutive features of 'politics as usual', gradually weaves in examples relating to the environment and aviation policy, and concludes, after a brief examination of the *Airports Commission*, that 'politics as usual' has been comprehensively reasserted both in the airports debate and in the environmental debate more generally.

WHAT IS 'POLITICS AS USUAL'?

Is the phrase 'politics as usual' just that: a convenient and cynical phrase? Well, it is a phrase and it sounds cynical: but I intend it to be rather more. In what follows I shall give a brief indication of what I take the phrase to imply by examining the political framework within which debates around aviation expansion are situated. In doing so I will try to give the phrase 'politics as usual' a reasonably precise meaning and show that it is exceedingly hard to break out from the default setting of 'politics as usual'. In particular, in relation to environment and aviation, the chapter seeks to encapsulate some of the challenges and points of contestation associated with national aviation policy.

Power and Presuppositions

What do I mean by the phrase 'politics as usual'? In broad policy terms, I mean the assertion or reassertion of certain standard approaches to the economy, to economic growth and development, to the protection of certain economic and political interests and a scepticism concerning the environment, especially where it is felt to be a brake on economic growth and business profitability. This does not necessarily imply a constant and active hostility to the environment, but it does imply a level of underlying indifference to its claims and a reactive rather than a proactive approach to their resolution.

However, 'politics as usual' goes deeper than the exhibition of particular policy commitments. In the sections that follow I will show that 'politics as usual' has a structure lying behind its particular manifestations. In short, this structure consists in a set of deep presuppositions, a lexical ordering between various levels of presuppositions and commitments, the structuring effects of power and influence and the limits of bureaucratic rationality. From this it follows that the visible actions of government in addressing environmental issues are an expression of underlying structures of power and influence, basic beliefs and administrative stasis. Midway between observable action and deep underlying presuppositions lie certain approaches to tactics and strategy, the manipulation and structuring of choices and the shaping of the political opportunity structure.

This can be illustrated by considering Steven Lukes's analysis of three dimensions of power to demonstrate the levels. At the deepest level (the third dimension), there is the power of unacknowledged action-guiding presuppositions which are non-observable, not questioned and never (or rarely) challenged (certainly not by anyone with the ability to make a difference). At the next level (second dimension), there is the ability to shape and manipulate agendas of public debate, through reputational power, or the implicit threat which arises (unstated) from the perception that powerful interests can access more explicit forms of power if they need to do so. And finally, there is the level of explicit observable power (first dimension). The key here is to recognise that power operates on all three dimensions and that power is not limited to observable manifestations of physical force or threat. These dimensions overlap and are not mutually exclusive; one of the primary differences between them is their relative efficiency. For instance, it is vastly more efficient to employ reputational power to manipulate the political agenda than to rely on explicit threats or the exercise of physical force; and if the underlying background to the political agenda is already structured by deep underlying unquestioned or naturalised presuppositions ensuring prior agreement on fundamental ends that is more efficient still (Lukes, 2005).[1]

For example, a political actor, if known (or believed) to be powerful, can rely on reputation and implicit threats to achieve success in agenda manipulation. Further, their power might draw additional nourishment from the underlying set of beliefs and presuppositions which absolve them of the need to argue their case explicitly precisely because their case is always already the default position.[2] This is a familiar state of affairs in which it is only challengers to the status quo who have to seriously argue their case (often in the face of ridicule) whereas defenders of the status quo are required to provide only minimal support for theirs, because it is seen, at some level, as beyond the need for argument. Thus, defenders of the political and economic status quo invariably start with an inbuilt advantage either because no one questions the desirability of economic growth (as an end), or of roads and planes and trade and industry (as means to that end); or, if these are questioned, the questioning is at best superficial and the challenger presumed to be defending something easily caricatured as 'unrealistic', or based on 'sentiment' or 'intuition.' To be real is to be measurable; whatever is not measurable is not real; and the approved medium of measurement is GDP. In such a debate – heard a thousand times on the TV and radio – it is inevitable that, at some point, the protester will have to assent to the desirability of economic growth, in however qualified a form. Once such minimal assent is wrung out of the objector, it is but an easy step to the conclusion that although there might be some hypothetical cases where economic considerations are not paramount, in practice unlimited economic growth (as standardly measured) is a desideratum beyond questioning or doubt. At this point, all that remains is discussion of means or location or timing: when or where, rather than whether; more or less, rather than none. And neat opportunities for the manipulation of decisions through constrained choices neatly open up. Here we might recall the decision making over the extension of the M3 through Twyford Down, Winchester, in the 1990s, or the current debate over the expansion of airport capacity, to which we shall return.

Let us return to the structure of 'politics as usual' and expand the point concerning presuppositions. 'Politics as usual' is about adherence – despite surface disagreement – to a set of presuppositions. It is possible to distinguish deeper presuppositions (assumed without argument and taken for granted in all political and economic argument) from surface presuppositions which are relatively shallow and (relatively) open to question and challenge. These surface presuppositions often (but not invariably) concern means rather than ends, just as deeper presuppositions tend to concern ends and not means. This is not, however, an absolute distinction: some very deep presuppositions in political ideology and doctrine primarily concern means.

The important point here is that not only are there action-guiding presuppositions operating at different levels, but there is typically a lexical principle governing the relations between them in cases where they might otherwise conflict. Borrowing from John Rawls, lexical ordering is akin to the rules governing the order of entries in a dictionary: words beginning with *a* precede those beginning with *b*, and *ba* precedes *bb*, and so on. In general, it is ordering governed by the principle that certain conditions have to be satisfied before others come into play (Rawls, 1999, pp. 37–38).

In a football league table, the primary ordering principle is the aggregate number of points a team has; the secondary ordering principle, which comes into play if two teams have the same number, consists in comparison of their respective goal difference. Rawls was, of course, concerned with justice, not soccer. His lexical ordering governed conflict in practice between the principles of justice he argues for within his theory. These are (a) the liberty principle and (b) the equality principle, the latter consisting of fair equality of opportunity and the difference principle (which justifies departures from equality if, and only if, to do so is to the greatest benefit of the least advantaged). For Rawls, in cases of conflict, the liberty principle should be satisfied first, then fair equality of opportunity and finally the difference principle (Rawls, 1999, p. 266). Overall, this suggests that, other things being equal, for Rawls, liberty is paramount and equality secondary; but a society which operated on Rawlsian principles would clearly be more egalitarian than one which did not.

For 'politics as usual', it would seem that once the conditions for ensuring economic growth are satisfied, environmental considerations can come into play, but not vice versa: environmental concerns do not have to be satisfied before economic concerns can be addressed.[3] Thus, if environmental protection is at the expense of economic growth it should not be pursued; if it positively promotes economic growth (as it does for optimistic ecological modernisers) it should be positively promoted; if it has no discernible effect on economic growth a degree of latitude appears in which it might or might not be pursued according to political preference.

Debates surrounding transport, railways, roads, aviation policy in general, or the taxation of aviation fuel, airport expansion, addressing the growth in air travel, including aviation within the emissions trading scheme, all take place within this framework. And it has now become clear that two (at least) of the presuppositions of 'politics as usual' are, firstly, the presupposition of the uncontested desirability of economic growth (as graphically revealed by fear of its opposite, a double dip recession); and, secondly, the presupposition that the relationship between the environment and the economy is typically a zero sum game. A variant on the second presupposition (more accurately, an unpacking of it in terms of lexical priority) is that environment and economy do not necessarily clash, but if and where they do, economy should be prioritised; and that although it might be possible for both to coexist in a green economy pursuing sustainable green economic development, the implicit proviso is that this holds only for so far and for so long as there is no clash. It is also worth remembering that green growth, the green economy and so on, are typically promoted not as green qua green, but as good for the economy[4]: in other words, environmental values are subordinated to economic values. As Bertolt Brecht might say, it's still 'Grub first, then ethics'.[5]

'Politics as usual' is, of course, constituted by other presuppositions, including beliefs concerning the proper scope and limits of governmental action, taxation, property ownership and rights, the status of (and response to) economic and social inequality. Here my aim is not to enumerate all of the presuppositions, but only those directly bearing on the topic of this chapter. However, before proceeding I wish to comment a little further on the presupposition of the value of economic growth and also to provide an example of a different sort of absolute presupposition, one not obviously related to the environment. To take the first: it should certainly be acknowledged that there are legitimate debates over the best means to secure economic growth. These debates are typically characterised by two things: discussion of side-constraints and of timescale. The two are intrinsically linked. By side-constraints I mean that, for some, providing or maintaining the conditions for economic growth and development should not come at the expense (say) of the underprivileged or unemployed (or those likely to lose their jobs as a result of structural change). This is a limitation on the immediate means to be employed in the service of economic growth. We can easily imagine the response about present sacrifice being necessary to ensure future prosperity, including (presumably) of those currently making the sacrifice. The point remains, however, that some maintain that immediate means which require this sort of sacrifice should not be employed, especially as the promise of future prosperity is an inadequate substitute (even if genuinely forthcoming) for suffering now. The second point concerns the timescale in which economic growth and prosperity is to be attained and sustained. Here one has to distinguish between what is required in the short and in the long term. To some observers it might appear that, at present, the British government do not *want* economic growth, and that through the fiscal means they are employing, they are doing all in their power to prevent it. The Government's reply would presumably be that long-term economic growth and development is conditional upon economic and financial restructuring in the present, and that putting the economy on a 'sound footing' is the only way to secure long-term prosperity and economic growth. The underlying presupposition of the unmitigated benefit of economic growth remains unchallenged. Again, in relation to the discussion of airport capacity in the United Kingdom, as typified by the debate of the third runway at Heathrow, the discourse centres on the notion of appropriate infrastructure for long-term prosperity and growth. Hence one can see governments playing a long game, allowing time for options to be aired and scrutinised. Cynically, of course, one might say that the presupposition is an unexamined one of continued economic growth tightly linked to a predict and provide model of airport capacity provision, ignoring environmental impacts other than local concerns about noise. Again, one might say that alternatives are left to collapse under their own weight while Heathrow calmly waits for a positive answer. More of this later. The point is that there is long-term thinking and there is long term thinking: the first type is simply the present, with its assumptions and structures projected in to the future; the second requires examination of and challenge to dominant assumptions and structures.

As an example of a presupposition of a very general nature, but which nonetheless bears on environmental policy, let us consider attitudes to the scope and limits of governmental action. In 2007 the political editor of the *Daily Express*, Macer Hall, remarked, in response to Zac Goldsmith[6] that:

> Goldsmith seeks to remove the ethical dimension – the exercise of free choice – from daily household decisions. Tories of his ilk seem to have stolen from socialism the idea that the tax system should be used to change human behaviour rather than being a necessary evil for funding essentials such as defence and policing. (Hall, 2007)

Hall is clearly right that the point of green taxation is its ability to change behaviour, and he is eloquently expressing a view traceable at least as far back as John Locke. This view would categorically rule out the use of green taxes on the grounds that the proper role of government does not (and should not) include behaviour modification through taxation. Taxation should be limited to provision of the basic public goods which are essential to the role of government as the provider of order for its citizens. On this view, green taxation and the proper use of taxation are mutually exclusive, and whatever environmental protection might be required, nothing could legitimately sanction green taxes as a means to securing it. This, then, is an example of a basic presupposition covering means rather than ends. Where presuppositions of this sort operate, even where the ends are deemed good, certain means are regarded as impermissible.[7] One could argue against the view expressed by Macer Hall that *in fact* it does not hold, and that in practice governments do not restrict themselves in the way he desires. The reply is that of course they do not, but Hall is stating a normative, action-guiding presupposition which indicates a direction of travel rather than a point of arrival, and which is no more refuted by facts than people driving too fast refutes the speed limit. Although adherence to this presupposition might be more honoured in the breach than in the observance, this does not prevent its invocation in arguments over the proper limits of government action.

Institutional Inertia

All policy making tends to be conducted incrementally, in accordance with the notion of bounded rationality. Disjointed incrementalism, as outlined by Charles Lindblom and others, is both descriptive and prescriptive. It describes how things happen and prescribes a model for rational decision making in complex societies. In the political issue attention cycle, environmental issues, as they wax and wane in the public and political consciousness when they return to active policy consideration, start not from scratch but from an already institutionalised footing. They gradually become institutionally embedded and operate within the bounds of the structural presuppositions of bureaucracy as an entrenched dimension of their policy making activity. Institutionalisation, in the form of appropriate agencies and departments, possesses the considerable advantage of providing a ready-made structure within which policy responses can be channelled. However, this can also be a drawback, because not only do governments and departments inherit methods and techniques for dealing with issues, but also problems tend to be defined in ways which admit of solutions within the prevailing political and administrative arrangements. Drawing on Herbert Simon's idea of bounded rationality, Douglas Torgerson argues that there are limits to, what he terms, the 'administrative mind' (2005). In contemporary societies this means that policy makers typically pay attention only to problems which are amenable to technological and administrative solutions. Modern 'rational' administration presupposes a vision of order and progress within which certain approaches or responses are regarded as reasonable or rational and others are not. Thus environmental problems are often responded to atomistically rather than holistically; to see them as interconnected would constitute a challenge to our underlying views of economic development. For Torgerson, the administrative mind cannot admit that there might be a fundamental flaw in the whole pattern of industrial development. Environmentalism frequently mounts a direct challenge to the administrative mind in that it asserts or implies that progress as commonly understood may be a source of disorder rather than order, 'disrupting the natural systems upon which civilization and human life depend' (2005, p. 105). The administrative mind accordingly seeks to denigrate those who articulate a vision different from the administrative norm:

> To be dealt with, the 'crisis' had to be viewed and treated not comprehensively, as the product of a basic flaw in the whole project of industrialization, but in a manner which identified *manageable* problems. Although the problems could be regarded as somehow

> commonly 'environmental', they had to be defined, in operational terms, as primarily separate, capable of being solved in a manner which matched the functional differentiation of the administrative apparatus. (2005, p. 106)

Fragmentation of issues into different government departments, each with its own logic and preferred policy instruments, means that the interconnected nature of many environmental problems is overlooked. On the one hand, such fragmentation is hard to avoid and there is obviously something to be said for reframing problems so that they admit of something approaching a solution. Postponing the solution to anything until we have the solution to everything is not a solution to anything. To that extent there is much to be said for incrementalism, bounded rationality and the limited (but practicable) constraints of the administrative mind. However, there should be limits to these limits. Systematically reframing problems to fit the solutions available is a modern, non-mythical, variant of the mistaken policy of Procrustes, who did not, it should be noted, *really* solve his bedding problem (occasioned by some visitors being too short and others too long for his bed) by either cutting off their extremities or stretching them to fit.

On a more general point, the workings of the administrative mind and its associated incrementalism are examples of how the structure of presuppositions, constitutive of 'politics as usual', are maintained and perpetuated through institutional structures. This demonstrates also why it is so hard to challenge, still less to modify or to remove them. Presuppositions embedded in forms of institutional life are powerful indeed.[8]

VIRTÙ AND *FORTUNA* IN ENVIRONMENTAL POLITICS

An important part of 'politics as usual' was characterised by Niccolò Machiavelli in the early 16th century as consisting in the interplay of *fortuna* and *virtù*. In direct descent are phrases such as 'a week is a long time in politics' (Harold Wilson) or 'events, dear boy, events' (attributed to Harold Macmillan). The relation between *fortuna* and *virtù* is the interplay between chance and abilities within the political opportunities presented to (or created by) the political agent. The term *virtuoso* perhaps better captures the spirit of Machiavelli's concept of *virtù* than does the standard English translation of 'virtue'. For Machiavelli, it refers to the personal qualities the prince should acquire to maintain the state and achieve great things.

The interplay of *virtù* and *fortuna* is an ineliminable presupposition of political activity. Obtaining office does not guarantee political success; but if a powerful politician – a virtuoso – holds a certain post they can achieve great things; holding the post is a necessary but not sufficient condition; *fortuna* and their *virtù*, their ability to grasp the opportunities that chance offers them, are necessary too.

Examples are ubiquitous, but we can consider some recent ones directly within the field of environmental policy. Upon being appointed Secretary of State for Energy and Climate Change in May 2010, Chris Huhne was immediately seen as pivotal to converting the enhanced profile of environmental concerns into successful policy. He was, as commentators say, a 'big beast' (one of the few Liberal Democrat 'big beasts') recognised as such both in professional party and parliamentary politics and by the public. Although it is a truth universally acknowledged that the Chancellor of the Exchequer is, *ex officio*, one the biggest beasts in the political jungle, Huhne was sufficiently powerful and *virtuoso* to harry the Chancellor and the Treasury in the discussions leading up to the formation of the Green Investment Bank. He did not fully succeed, but that he succeeded at all is the point (see Connelly, 2011). Huhne was also an important player in international climate change negotiations, as a forceful negotiator.[9] However, in other realms his *virtù* proved limited and *fortuna* decreed that he would be forced to resign in February 2012. His successor, Edward Davey (another Liberal Democrat) is regarded as both less green and less *virtuoso* than Huhne. Two points should be noted here. The first is that Liberal Democrats do not, generally speaking, lead the large and powerful departments and therefore have to be possessed of extraordinary *virtù* to be able to challenge the Treasury, the other big ministries, and the deeply held presuppositions of their ministers and civil servants as expressed in their words, thoughts and deeds. Liberal Democrat 'big beasts' have to be bigger and more beastly than those already occupying the major offices of state. Unlike the Chancellor, they have to achieve greatness rather than have it thrust upon them.

Some other examples. Before 2010, Justine Greening, the MP for Putney, was known as an eloquent and vociferous opponent of a third runway at Heathrow Airport. Given that her constituency is directly under the flight path, it could be argued that hers was essentially a Not In My Back Yard-inspired (NIMBY) opposition. Be that as it may, it nonetheless fitted neatly into the Conservative opposition's objections to Labour's plans for airport expansion. Following the election, after a brief spell as Economic Secretary to the Treasury, she became Secretary of State for Transport in

October 2010. This was interpreted as a positive symbolic move, given her known opposition to Heathrow expansion. However, her removal in September 2012, when she was appointed (in the eyes of many, demoted to) Secretary of State for International Development, was the mirror image of the same symbolism. She was replaced as Secretary of State for Transport by Patrick McLoughlin who apparently has a fear of flying, but not necessarily of airport expansion. Greening's deputy at the Department for Transport, Theresa Villiers, was opposed to the expansion of Heathrow and to London Mayor Boris Johnson's proposal for a new airport in the Thames estuary; she was also in favour both of High Speed 2 (HS2, the successor to HS1 *Eurostar*) and freeing capacity at Heathrow through increased use of Manchester and other northern airports.

It is worth reflecting for a moment on the possibilities of leadership on a given issue for a minister. Leadership is a complex topic with definitions abounding. The key variables in effective leadership are power or its absence and scope or its absence. For example, a powerful actor, such as the Treasury, can dominate without the need to persuade; the relatively powerless, on the other hand, can lead through skill in negotiating or facilitating agreements, or in redefining and conceptualising the terms of debate. Again, some leaders seek incremental change only whereas others seek – and might achieve – transformation. A Minister in the Department of Transport or Environment is likely to be relatively powerless and hence likely to display entrepreneurial or cognitive leadership types. Whether they succeed in being more than incremental depends on a confluence of circumstances, although tactically their leadership style is likely to be humdrum rather than heroic or transactional rather than transformational.[10] In the longer term, transformation can occur through cognitive leadership, to that extent 'politics as usual' can be subverted, but in the short term, the hegemony of powerful actors expressed both through their hard power and their reliance on implicitly agreed presuppositions will tend to dominate.

When Justine Greening left transport, it was reported that there was rejoicing in the aviation industry at the removal of an opponent. Elsewhere there was condemnation of the move and suspicion of its motives. Boris Johnson claimed that 'there can be only one reason to move her – and that is to expand Heathrow Airport'. He proclaimed that:

> we will fight this all the way. Even if a third runway was built, it would not do the job of meeting Britain's needs … If we are to remain Europe's premier business hub we need a new four-runway airport, preferably to the east of London, that addresses the problem of aviation capacity before it is too late, and business is driven into the arms of our European competitors … It is time for Patrick McLoughlin to look at all the

options, including bolder solutions that would deliver massive benefits in jobs and growth. (BBC News, 2012)

But make no mistake. Johnson may be opposed to expansion at Heathrow, but is most definitely not opposed to airport expansion as such, which he positively promotes. Zac Goldsmith observed that with Greening's appointment to Transport, the Prime Minister's position on Heathrow was solid and that yielding so easily suggested 'panic, not principle',[11] while *Friends of the Earth* criticised the replacement and claimed that Greening had been 'shunted out' and was also a 'victim of intense aviation lobbying over airport expansion (BBC News, 2012).

Meanwhile, at Department for Environment, Food and Rural Affairs (DEFRA), in September 2012, Caroline Spelman was replaced by Owen Paterson, who might or might not be a climate change sceptic (he has signally failed to clarify his view) but who is certainly not regarded as an environmental activist (Carrington, 2012; Monbiot, 2012). Prior to his new appointment, while still Northern Ireland secretary, he set out a three-point plan for economic growth in which the key elements were the exemption of all micro businesses from red tape, the ending of all energy subsidies and fast-track exploitation of shale gas, together with the 'urgent review of airport policy to ensure Britain gets its full share of global trade' (ConservativeHome, 2012). His appointment cannot be seen as a sign of continued environmental commitment by the government.

The precise difference these personnel changes will make is a matter for conjecture: but powerful environmental voices who might (given their *virtu* and good fortune) at least have nibbled at the edges of 'politics as usual' have been removed from positions of influence. In their absence, it is likely to remain unchallenged in the face of the powerful forces of the treasury and its allies. These days David Cameron is a largely silent figure on the environment, having conceded virtually everything to the consolidated forces of economic orthodoxy, otherwise known as 'politics as usual'.

TRAINS AND BOATS AND PLANES

I will continue with some reflections on where this all leaves us. What we have seen is government returning to type. That it is a predominantly Conservative government is important – perhaps it was more willing to return to type than governments of another stripe – but we should recall the extent to which the presuppositions of 'politics as usual' are shared by all mainstream parties and the political class as a whole.[12] It might be argued that the decision to proceed with HS2 is an exception. The counter argument is to point out, firstly, that competing with air travel uses fuel and generates emissions comparable with aviation; secondly, that the majority of environment groups (not just NIMBY groups, but Greenpeace, Friends of the Earth, the RSPB and others) are opposed to it; thirdly, that the arguments propounded in its favour have not, for the most part, been couched in environmental terms, but almost entirely in terms of economic growth and development; and, fourthly, that if the intention is to improve the rail system overall, the best way to do so is probably to spend it on upgrading, lengthening trains and platforms and re-opening lines. Support for HS2, whatever its merits, is not in itself a sign of environmental commitment.

A quick comment on NIMBY, which is typically used as a term of disapprobation for a narrowly self-referential and selfish frame of mind indifferent to the common good, where it infringes the protesters' perceived private local good. If it were merely and always that, this would be a justified charge. However, experience shows that NIMBYism can be transformative and transcend its starting point, as in the case of protest against road expansion in the 1990s when local protests joined together in national protests against the national roads programme. NIMBY can, under certain conditions, be transformed into NIABY – 'not in anybody's backyard.' As individuals become radicalised locally they tend to start considering the broader implications of the project and their opposition to it; ultimately they are drawn to take a stand encompassing all aspects of the national transport and environmental debate. This is not inevitable, but it is a definite and discernible tendency, in protests against road building, HS2 and airport expansion, especially where local groups form working alliances with national and international campaigning environmental groups such as *Greenpeace*, *Friends of the Earth* and *Plane Stupid.*

THE CONTRACTION OF THE AIRPORT EXPANSION DEBATE

Over the past couple of years, chance and circumstance have combined in a nice easy shuffle towards setting up a decision in favour of a third runway. This claim is based on the view that expansion is going to happen (is seen as necessary for competitive economic reasons), that Boris's Island was essentially a decoy which deflected attention from *whether* expansion was desirable to *where* it was desirable; that, Boris's Island being shown to be impossible, only one practicable choice remains – the expansion of Heathrow.[13] In July 2013 Heathrow Airport submitted its preferred third runway options to the Airports Commission. The BBC transport correspondent commented that in 2012 he had been told by government officials that the prospect of building a third runway at Heathrow:

> was 'dead and buried' and [that] they'd look at 'all ideas bar a third runway'. Back then we also had a Transport Secretary, Justine Greening, who lived under the flight path and had campaigned against the plan. What a difference to today, when a third runway at the UK's biggest and busiest airport is very much back on the table. (Richard Westcott, BBC News, 2013)

'Politics as usual' is thereby reasserted, with the overriding presupposition of the primacy of economic growth and development to the fore, and the tactics of a constrained choice between one evil and another (perhaps slightly) lesser evil; and, to top it all, is a side dish of environmental concern: after all, a new airport in the Thames Estuary would be environmentally very destructive, is opposed by all the environmental groups and therefore, on environmental grounds, Heathrow is the least worst option. This all amounts to a textbook masterpiece of political manoeuvring and represents the power and *virtù* of the Chancellor and his allies within the government and their ability to grasp opportunities when they had them. George Osborne has increasingly revealed his support for more runway capacity in the south-east of England and for Heathrow as the only practicable solution, although still claiming that all options should be open. That is, all options are open, except *not* increasing capacity; and although all options are open, some are more open than others. Against this background, it seems unlikely that there will be any systematic discussion of aviation in relation to transport policy as a whole or in relation to climate change emissions and the United Kingdom's targets. In so far as emissions are considered, it seems to be that air expansion is assumed and the necessary cuts will have to be made elsewhere.

Is this conclusion based merely on a hunch and a cynical misreading of the political runes? I submit that the final piece of evidence is now to hand. Although the government has continued to claim that its position against airport expansion remains unchanged, this seems rather implausible in the face of the appointment of the Airports Commission, under Sir Howard Davies, as the central plank in its consultation process on future airport capacity.

The Airports Commission

On 2 November 2012, the Secretary of State for Transport, Patrick McLoughlin announced the terms of reference for *The Airports*

Commission (McLoughlin, 2012). It should report no later than the end of 2013 on:

- its assessment of the evidence on the nature, scale and timing of the steps needed to maintain the United Kingdom's global hub status and
- its recommendation(s) for immediate actions to improve the use of existing runway capacity in the next five years – consistent with credible long-term options.

And it should report no later than summer 2015 on:

- its assessment of the options for meeting the United Kingdom's international connectivity needs, including their economic, social and environmental impact;
- its recommendation(s) for the optimum approach to meeting any needs and
- its recommendation(s) for ensuring that the need is met as expeditiously as practicable within the required timescale.

In launching the Commission, McLoughlin remarked that 'Aviation is vital to the UK economy and we need to have a long-term aviation policy which meets the challenges of the future.' The presupposition is that there will be an increase in airport capacity: the Commission is asked to deliberate 'on the nature, scale and timing of the steps needed to maintain the United Kingdom's global hub status'. This is not a policy debate about the future of aviation so much as a debate about the location and expansion of aviation for the future. The timescale, it should be observed, is politically expedient, neatly allowing the government to uphold the letter of the coalition agreement which ruled out airport expansion in the current parliament.[14]

Vocal Green to Silent Green to Ungreen

Since its election the government has moved decisively away from its loudly proclaimed green commitments, through a phase of relative silence, to a reassertion of traditional (ungreen) values. The most recent indication of this can be seen in the Prime Minister's recent speech on the need to renegotiate the EU's powers to dictate areas of environmental policy if the Conservative Party wins the next election. He claimed that there were 'areas – social legislation, employment legislation, environmental legislation – where Europe has gone far too far.' He was not specific on details, but close political observers think that the Conservative Party would, while continuing to accept the EU carbon emissions reduction target, oppose the setting of a separate renewables target; that they would act in the light of Osborne's long established complaints about the 'ridiculous cost' the Habitats and Birds Directives place on UK businesses (despite a DEFRA investigation showing the falsity of this claim); and that they would seek to make changes to the Air Quality Directive, where the United Kingdom has continually failed to meet its standards (ENDS, 2013).

CONCLUSION

My not very well-kept secret is that there has been no descent to 'politics as usual', rather, there was an apparent ascent from 'politics as usual' which masked the underlying normality, the smoothly working deep presuppositions of the political system. When the illusion of ascent is removed, the observer mistakes the process for descent, whereas it is a revelation of the level at which 'politics as usual' was always operating. For the advocate of 'politics as usual', everything is neatly back in its rightful place. The Treasury reigns supreme; radical, green and powerful ministers prepared or able to challenge the Treasury have resigned or been sidelined and replaced with weaker or actively hostile ministers; public opinion is indifferent and believes the claims of green government to be bogus; the deep lying assumption that economic growth is paramount is triumphant. Gestures are made to green growth and sustainable development, and to the politics of ecological modernisation, but they are trumped where necessary by standard economic concerns. Hypothetical countervailing instances are weak or misleading, for example HS2, which sits far more snugly within the standard growth and economic development paradigm than the green agenda. There are of course exceptions, and where they exist they should be celebrated; but the juggernaut of politics as usual continues relatively unimpeded by carrying the occasional passenger looking in a different direction.

Of course, this is to some extent an exaggeration. But it is certainly true that public concern with, for example, climate change, does not run in parallel with the gravity of the issue – if anything, interest and concern has declined in many countries. Again, far from adopting the view that environment and the economy can be happily coexisting bedfellows, even at the rhetorical level, the general public tends to still regard them as firmly opposed one to the other.

Should my conclusion be thought to be unremittingly negative and, indeed, verging on the cynicism I disparaged in my introduction, I would reply that change is possible although it is typically a lot slower than people want it to be; and that large events, although not determining the response to them, nonetheless create a space in which a response is required. In this sense the evidence of the effects of climate change can and probably will require at some point that certain of our deep underlying presuppositions be revalued and modified, with consequent changes in our actions. But I have argued in the foregoing that modifying or dislodging the presuppositions and other features constitutive of 'politics as usual' is at the very least a Herculean and possibly a Sisyphean task.

NOTES

1. This account is indebted to (but not intended to be an accurate account of) Lukes. By 'naturalised' I mean those occasions where certain beliefs are so much taken for granted at a certain time that not only are they never questioned, but are simply taken to be the natural order of things. For a short account, in the context of environmental policy, see Connelly, Smith, Benson, & Sanders, 2012, pp. 156–62.

2. This point is independent of the objective truth or falsity of the claim itself: it concerns the likelihood of believing or accepting something independently of the purely rational reasons for assenting to its truth.

3. It might also be noted that deciding on *whether* a principle has been satisfied is a contentious and contested matter. For instance, one can readily imagine that, depending on its interpretation, the principle of maximizing the conditions for economic growth might never as satisfied.

4. A point made by commentators seeking to draw the Chancellor's attention to the fact that the green economy is currently growing whereas the conventional economy is not. It would appear that the current Chancellor of the Exchequer, George Osborne, is so opposed to environmental concerns that he is inclined to forget that there is a green contribution to the economy at all.

5. The ideas expressed in the foregoing passages owe something to Collingwood 1998 [1940]. For further development of some of these points in an environmental context, see Connelly et al., 2012). Many of the points could, of course, be re-expressed, *mutatis mutandis*, in the language of ideology or in the Marxist idiom.

6. Then an adviser to David Cameron and later to become a Conservative MP.

7. As a matter of fact, many holding this view would hold the ends to be undesirable too; this leads to the result that certain means are held to be impermissible both because they are means to improper ends, and also because they are improper means irrespective of ends.

8. For further discussion of the topics in this section, see Connelly et al., 2012.

9. A point recognised by the Greenpeace UK Director, John Sauven, in responding to the news that Huhne would remain at the 2010 Cancun climate

change negotiations and not return to London for a parliamentary vote. Sauven said that Huhne had a vital role in moving the talks forward: 'with the outcome of the Cancun summit now hanging in the balance, attention must now focus on whether he can help to broker a climate deal' (BBC News, 2010).

10. For a definition and typology of types and styles of leadership, see Wurzel and Connelly (2011, p. 13).

11. It should be noted that Goldsmith would regard a *volte face* on Heathrow expansion as an act of betrayal which would lead to his resignation as a Conservative MP. Justine Greening, equally, would very likely continue her opposition by resigning from the cabinet.

12. This is an example of Lukes's third dimension of power.

13. For a similar view, see Carrington, 2012b.

14. The history of decision making on airport expansion is an example of agenda setting and corresponds to Lukes's second dimension of power. Whether the Conservatives will in fact be in a position to take advantage of the commission's recommendations is a moot point. The chances of the Conservative Party winning a single party majority in the House of Commons in the 2015 election have been significantly diminished by the recent vote against their proposals to reduce the number of, and equalize the numerical size of, parliamentary constituencies.

REFERENCES

BBC News. (2010, 4 December). *Chris Huhne to stay in Mexico for UN climate talks.* Retrieved from http://www.bbc.co.uk/news/science-environment-11950882. Accessed on 12 February 2013.

BBC News. (2012, 4 September). *Boris Johnson's Heathrow warning after Justine Greening's move.* Retrieved from http://www.bbc.co.uk/news/uk-politics-19475249. Accessed on 10 February 2013.

BBC News. (2013, 17 July). *Heathrow submits third runway options to Davies Commission.* Retrieved from http://www.bbc.co.uk/news/uk-23337754. Accessed on 17 July 2013.

Carrington, D. (2012). Owen Paterson's climate change problem: cock-up or conspiracy? *The Guardian*, 7 September 2012. Retrieved from http://www.guardian.co.uk/environment/damian-carrington-blog/2012/sep/07/owen-paterson-climate-change-sceptic. Accessed on 20 February 2013.

Collingwood, R. G. (1998 [1940]). *An essay on metaphysics* (rev. ed.), edited with an introduction by Rex Martin. Oxford: Oxford Clarendon Press.

Connelly, J. (2009). Voting blue, going green? David Cameron and the environment. In M. Beech & S. Lee (Eds.), *The Conservatives under David Cameron.* Basingstoke: Palgrave Macmillan.

Connelly, J. (2011). Vote blue, go green, what's a bit of yellow in between? In M. Beech & S. Lee (Eds.), *The coalition government.* Basingstoke: Palgrave Macmillan.

Connelly, J., Smith, G., Benson, D., & Sanders, C. (2012). *Politics and the environment: From theory to practice.* London: Routledge.

ConservativeHome. (2012, 16 May). Retrieved from http://conservativehome.blogs.com/thetorydiary/2012/05/unemployment-is-down-exports-are-booming-but-paterson-hague-and-fox-all-urge-chancellor-to-do-more.html. Accessed on 20 February 2012.

ENDS. (2013, 24 January). Europe UK PM calls for roll-back of EU green laws. Retrieved from http://www.endseurope.com/30419/uk-pm-calls-for-rollback-of-eu-green-laws?referrer=bulletin&DCMP=EMC-ENDS-EUROPE-DAILY. Accessed on 20 February 2012.

Hall, M. (2007). How dare millionaire Goldsmith lecture the voters on "sacrifice". *The Daily Express*, 12 September 2007.

Lukes, S. (2005 [1974]). *Power: A radical view* (2nd Rev. ed.). Basingstoke: Palgrave Macmillan.

Monbiot, G. (2012, September 6). *Declaring war on the environment*. Retrieved from http://www.monbiot.com/2012/09/06/declaring-war-on-the-environment/. Accessed on 20 February 2012.

Rawls, J. (1999 [1971]). *A theory of justice* (Rev. ed.). Cambridge, MA: Harvard University Press.

Torgerson, D. (2005). The ambivalence of discourse: Beyond the administrative mind? In R. Paehlke, D., & Torgerson (Eds.), *Managing Leviathan: Environmental politics and the administrative state* (2nd ed.), Peterborough, Ontario: Broadview Press.

Wurzel, R., & Connelly, J. (2011). *The European Union as a leader in international climate change politics*. London: Routledge.

CHAPTER 11

THE DEVELOPMENT OF FRANKFURT/MAIN AIRPORT: A TRADITIONAL NARRATIVE OF LOSS AND GAIN

Ute Knippenberger

ABSTRACT

Purpose – *This chapter focuses on the conflicts that have accompanied the development of Germany's largest airport at Frankfurt/Main and examines how these conflicts are embedded in regional governance and airport planning. In the German context, the governance discourse is understood as a multi-faceted framework of actors, institutions and legislation that structures the actions of involved people.*

Method/approach – *The conflicts within this framework are structured by two aspects: the spatiality of the contested object, which results in tactical descriptions of the space involved, and an understanding of airports as parts of large technological systems which strongly influences the discourse of airport expansion. As spatial conflicts regarding large infrastructures develop over a long period of time, the chapter charts the historical evolution of the expansion conflict from the 1980s until today. It analyses the governance structure and the actor's constellations*

Sustainable Aviation Futures
Transport and Sustainability, Volume 4, 239–264

ISSN: 2044-9941/doi:10.1108/S2044-9941(2013)0000004011

throughout time to shed light on the use of narratives and the stability and change of pro and contra coalitions. It concludes with a critical reflection on the role of airports in the German planning system and the problems of governance in the multilevel, multi-actor environment of airports.

Originality/contribution – *The chapter highlights the necessity of examining the environmental impact of airports at different scales, from neighbouring cities to the large labour market stretching up to 100km from the airport.*

Keywords: Airport regional governance; spatial conflict; institutions and path dependencies

Frankfurt airport (FRA) is the third busiest passenger airport in Europe and the 11th busiest in the world handling around 57 million passengers a year. With around 2 million tonnes of cargo being handled annually, it is also among the 10 largest cargo airports in the world (Fraport AG, 2013). FRA is characterised by its hub activity and by its strong hinterland formed by the poly- or rather oligocentric Rhein-Main region, territorially located in the state of Hesse, one of the German *Länder*. It is operated by Fraport AG, a corporation which was floated on the German stock exchange in 2000. The airport is located within the so-called 'city forest' (*Stadtwald*) of Frankfurt on a six-hectare site mainly situated within the boundaries of the city of Frankfurt and in the middle of a densely populated region. Constant friction characterises the relationship between the airport and its surroundings. A long history of protest is associated with it and in the early 1980s, local unrest ultimately led to two policemen being shot and killed.

As a result of the opening of a third runway in October 2011, the region's politically astute and articulate elite (or *Leistungsträger*) are now affected by airport noise and the furore over expansion and airport operations has been revitalised. As aircraft noise is also now audible in the formerly almost unaffected city of Frankfurt, the protest has even resulted in a local Social Democratic Party (SPD) candidate being elected as the new city mayor, ending a 17-year reign of the Christian Democratic Union (CDU) under the leadership of the popular *Petra Roth*.

All this happened despite the fact that recent expansion plans were accompanied by a 10-year long mediation process and numerous public hearings. Crucially, one central outcome of this mediation, a night flight ban, was omitted in the final plan that was approved in 2007. This was perceived as a major betrayal of trust and might have fuelled the general opposition against the airport. By omitting a mediation outcome, decision makers have taken the risk of ramifications that were anticipated. The aim of this chapter is to examine the reasons why this occurred in the context of airport expansion governance and sustainable aviation policy in Germany.

AIRPORTS AND REGIONAL GOVERNANCE

The chapter will discuss the case of FRA under the notion of regional governance, framing the issue as an expression of a spatial conflict related to infrastructure. Regional governance is here understood to be a sum of all forms of collective regulation of social issues, from hierarchical steering to different forms of interaction between private and public actors (Mayntz, 2004) Especially among planners and planning-related sciences, governance has been adapted in the German discourse as a normative approach in the sense that governance meant the inclusion of the *Zivilgesellschaft* into a cooperative state, or that governance itself would be an expression of public participation (Einig, Grabher, Ibert, & Strubelt, 2005; Fürst, 2007). Instead of this, governance here looks at how collective issues are actually negotiated, irrespective of the form of negotiation that is used.

When we observe the governance of an airport in that sense, we are confronted with numerous actors and institutions on different governmental and societal levels that are in a closer or wider sense part of that negotiation. In order to make the material feasible to operate, a model of collective actors in different forms and levels of institutionalisation is used (after Mayntz & Scharpf, 1995; Scharpf, 2006). Scharpf characterises actors as structured by their institutional context and their acting is thus limited by their preferences, power and their perception with regard to the negotiated issue (see Fig. 1).

As an airport represents a concrete object with spatial relevance, preferences and perceptions of actors will be influenced by the spatiality of the object. Reuber (1999) points out that actors in spatial conflicts always carry an image of the space within the conflict and use this image as a strategy in negotiations. Especially in long-term escalations when the object of conflict itself is dynamic, its changing condition over time and history is important

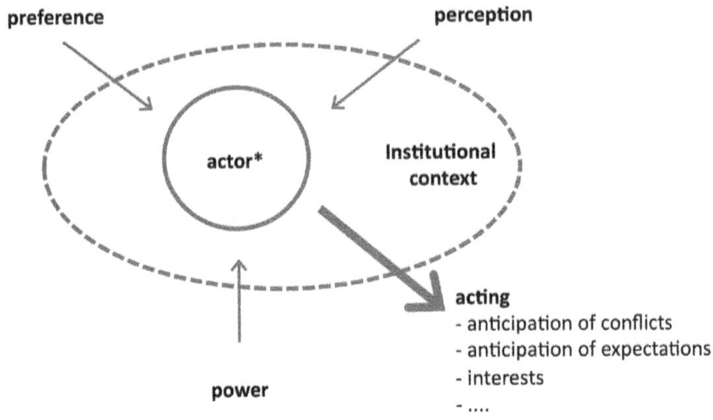

Fig. 1. Collective Actor Model Based On Scharpf (2006).

(*ibid.*). Therefore conflicts are often also structured by lock-in-effects and path-dependencies.

For the collective governance of airports this presents a complex task: not only are we confronted with a multiplicity of actors but also different scales of influence and impact. Airports are, by their very nature, globally connected. Of course, the international air traffic market plays a strong role in shaping the fortunes of an airport company just as international regulations structure the airport company's actions. Therefore, airports tend to show characteristics of Large Technical Systems (LTS) (see Hughes, 1987) and are to be understood as parts of the large subsystem of 'air-transport'. Following Mayntz (1993) such systems are characterised by a large network of technical and social components whose aim is to permanently achieve a certain purpose. Air-transport comprises a network of technological components, including airports, air-traffic control and air-craft, and has a high level of institutionalisation both through high-level organisations such as ICAO but also through its own modes of formal education and specialised degree programmes. Thus LTS tend to develop an intrinsic logic based on an instinct of self-preservation and their status and size also influence society. The intrinsic logic of a LTS means that it can benefit from information asymmetries within a local governance structure. Consequently, in order to refute the arguments of an airport operator, its opponents need to gather much knowledge.

Secondly, on the local level, airports have influence on very different scales that, at least in Germany, are transgressing municipal, county and sometimes national boundaries. The individual location of the airport and its territorial embededness structures airport governance. Different distances from the airport are coping with or profiting from the airport. Starting with the airport as a physical entity, it might be located within an eponymous municipality (like Frankfurt) or not (Amsterdam airport is within the municipality Haarlemmermeer). The surrounding cities or villages might be few or many. Airport noise might strongly affect some of them, others not at all, depending on the trajectory of flight paths rather than distance from the runway. They might also host many airport-related companies who require close proximity to the airport. The next level is the larger area of the airport workforce. It is estimated that Frankfurt has an employment catchment area of around 80 km (Langhagen-Rohrbach, 2002) and hosts companies who highly value the connectivity without being in a direct supply-chain to the airport, such as banks, insurance and consulting firms. But also noise might be a topic here. At all these scales, places and people might suffer or profit from the airport, sometimes at the same time.

Thirdly, as stated above, many state levels are part of airport governance, and complex bureaucratic procedures enlarge the number of actors involved. In Frankfurt, for example, to name the most important, it is the states of Hesse and Rhineland-Palatine, the region of South-Hesse and the state ministries of transport and environment, the metropolitan planning organisation of Rhine-Main, the cities of Frankfurt, Darmstadt, Mainz and Wiesbaden and many more municipalities in the closer and wider airport context and the national state, though in Germany the formal responsibility for approval procedures lies with the state level. Fig. 2 shows the state boundaries and the regional and metropolitan level.

THE GOVERNANCE OF AIRPORT EXPANSION IN GERMANY

This situation is producing actors with very different aims and perspectives. A highly differentiated system of planning approval in Germany is the result of the complexity of airport governance, with the three factors stated above, the existence of a LTS, the different scales of impact and the resulting multiplicity of actors. It combines a sectorial procedure aimed at the infrastructure in combination with a spatial planning procedure. In Germany, technical infrastructure is planned by an improvement procedure

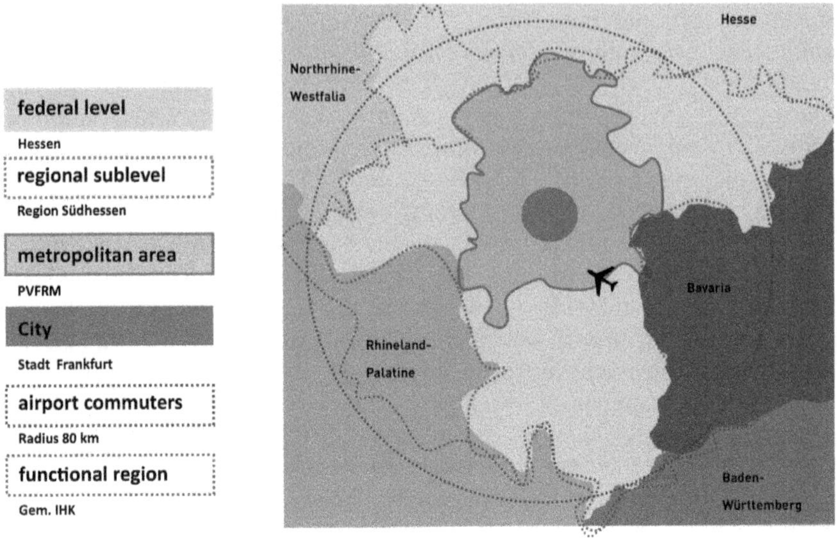

Fig. 2. Administrative Boundaries Within the Frankfurt Airport Region (Illustration by Author).

which runs in parallel with the hierarchical system of general planning, allowing infrastructure to be realised with powerful tools such as (albeit the rarely used) expropriation of private land. The relation to general planning is characterised by the *Fachplanungsvorrang*, the domination of sectorial planning upon general planning, often leading to friction between the different perspectives of infrastructural and integrated regional or urban planning.

Fig. 3 illustrates the German system. On the left are general or spatial planning with different state levels, seemingly hierarchic, but whose basis is the so-called *Gegenstromprinzip*, meaning that decisions are undertaken with the principle of countervailing influence. The sectorial planning is on a separate planning stream and ends with its own form of plan approval, the *Planfeststellungsbeschluß*. It becomes clear that the sectorial procedures are not anchored in the linear structure of spatial planning, since its implementation power relies on the primacy of the sector, a dominance that creates potential conflict (see Bickenbach, Kumkar, Sichelschmidt, Soltwedel &Wolf, 2005; Wolf, 2003; Ziekow, 2002).

For air traffic infrastructure, a policy document at the national level exists with the 'National Airport Concept', an informal document which

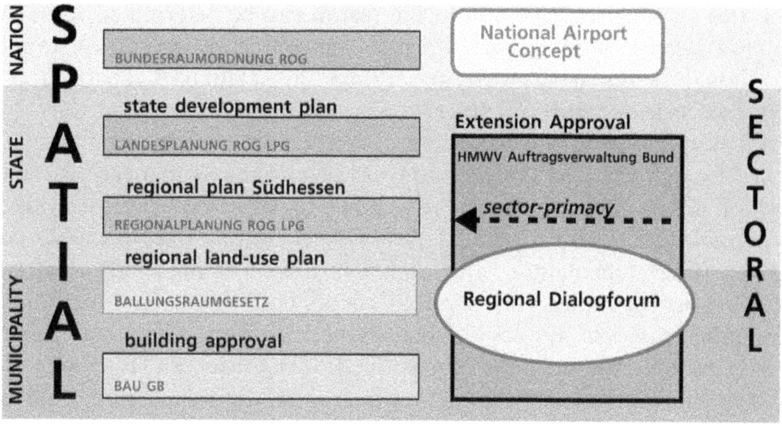

Fig. 3. German Planning System With Spatial and Sectoral Procedures (own illustration).

states the important role of air traffic for national and regional economies. Ecological side-effects of air traffic have to be taken into account while allowing air traffic to prosper (BMVBS, 2009). But planning is executed at the level of the German federal 'Länder'; for FRA, it is the state of Hesse. The Regional Dialogforum (RDF) shown in the oval in Fig. 3 is not part of the standard procedure but was part of the mediation procedure for the enlargement of FRA. It was an informal instrument whose role in the formal procedure was not clearly defined in the beginning.

BALANCING INTERESTS

A core aspect of all planning procedures in Germany is the possibility of formulating '*Anregungen und Bedenken*' – suggestions and concerns of third parties and the balancing of interests expressed there. The role of public and private goods is central here. Typically in an airport expansion context, the negative impact on private goods such as influence of noise on property prices would be balanced against the positive macroeconomic effects of increased air-transport capacity and enhanced connectivity. The latter could also be weighed against the negative impact on the public good of the natural environment. In the approval decision compensatory measures such as reforestation in the region can be determined to balance the negative ecological impacts on site. The compensation of private interests though is not part of the procedures, a fact that Bickenbach et al. (2005) see as a key reform demand for the system.

A key question in the balancing of interests regarding airport enlargement (through not new construction) however remains unanswered: which *level* of air-traffic-connectivity is producing significant positive macroeconomic impacts? Button and Lall (1999) as well as Button (2002) point out the positive economic effects of a hub airport on a city. But what are the positive indirect an induced economic effects of *more* hubbing? How much hub-traffic does the airport need to serve the public interests is a question that remains unanswered – at least in case one does not consider the goal of 'remaining Europe's third busiest hub-airport' sufficient evidence of public interest. This aspect, at least in the approval procedure for the enlargement of Frankfurt airport, has been put up by some suggestions and concerns, but has not been discussed in depth. The wording to not discuss these aspects in the text of the approval is simple: '*The interests of the applicant are coherent with the public interests*' (Hessisches Ministerium für Wirtschaft, Verkehr und Landesentwicklung, 2007a).

THE HISTORY OF FRA

The airport of Frankfurt/Main was established as a military airfield in 1936 and two aspects of its geographic location continue to influence airport politics today. Firstly, the chosen location was in the middle of a city forest in the south of Frankfurt/Main. Secondly, in 1944, the national socialist regime started with the construction of two new close parallel runways on an east-west alignment. The construction was started using forced female labour who were taken from Auschwitz and housed in a nearby concentration camp erected especially for the construction (Neff, 1997). When the jet engine became the propulsion mode of choice for post-war civilian aircraft, the original runways turned out to be too close to one other to permit simultaneous operations. Therefore, in the 1960s, plans to enlarge the airfield were proposed and the concept of a new runway, aligned north-south, and later known as *Startbahn 18 West* (Hessischer Rundfunk, n.d.), was formulated (Johnsen, 1996). Fig. 4 shows today's layout and planned projects until 2020.

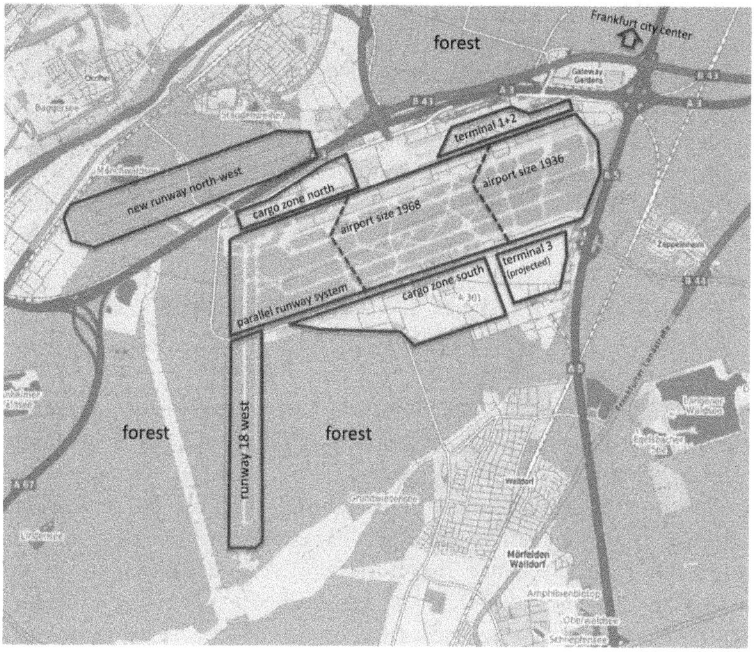

Fig. 4. Airport Runway Diagram and New Projects. *Source*: Albert Speer & Partner AS&P.

In the 1960s, public participation was a minor part of this procedure. When plans for the third runway became public and it became clear that many hectares of forest would have to be cut, a civil protest movement, in a climate of *Umweltbewegung*, arouse. The protest was not only limited to young activists, but included members of the middle and upper classes. The situation escalated in 1980 when the Hessian minister for economics, *Heinz-Herbert Karry*, gave the order for the immediate execution of the plan, meaning the construction of the new runway could commence before all court decisions had been taken. Contemporary witnesses have deemed this an unnecessary act of state power that was not even officially supported by the state minister of that time, Holger Börner. In 1981, Karry was assassinated in his private house – his murderers were never found and any clear connection to the protest movement against the runway remains unproven. His murder, and the deaths of two policemen during the anti-runway protests, are collectively referred to as the 'Startbahn deaths'.

The immediate commencement of runway construction works ran totally counter to any approach towards communicative strategies politics had started and destroyed any form of trust the protest movement might have had in official institutions (Johnsen, 1996). As a reaction, local politicians of all parties in the nearby town Mörfelden-Walldorf started a hunger strike. Nevertheless, construction continued and the new runway was inaugurated in 1984.

BEARING CONSEQUENCES AND DISCARDING THEM AGAIN

The conflict about Runway 18 West was not a locally confined story but one that had strong influences on German national politics. It is claimed to have taken *DIE GRÜNEN* (the greens) from mere eco-activists to state politicians and *Umweltpolitik* (environmental politics) from marginal to mainstream. It also became a paradigm for change from a paternalistic and authoritarian state towards a more cooperative and participatory style of politics (Sack, 2002). However, it also established a strong path dependency in airport-related politics in general and especially in Rhine-Main.

The memory of Runway 18 West was still fresh when, in 1997, the then Chairman of Lufthansa advocated in *Der Spiegel*, a weekly magazine, further airport enlargement (Der Spiegel, 1997). The then state government, a coalition between the SPD and the Green Party (DIE GRÜNEN), had stated in their coalition manifesto that the airport would be confined to its existing boundaries. The minister of state, Hans Eichel, therefore initiated the *Gesprächskreis Flughafen*, an expert Council with a range of different actors, including some from the former anti-runway movement, to establish an open dialogue about the claims – while at the same time trying to remain in a neutral position (Geis, 2003). The Council recommended a mediation procedure which took place from 1998–2000. The outcomes was that an expansion of the airport was necessary owing to its major influence on the regional and national economy, but that any development had to embrace five issues (Hänsch, Niethammer, & Oeser, 2000):

- Optimisation of the existing airport and runway system
- Capacity enlargement by physical development
- A night flight ban
- An anti-noise package
- A regional dialogue forum (RDF)

The RDF here took an overarching role in order to manage the other four issues and to establish an on-going dialogue about alternatives and their consequences. This was to enable the development of a 'No-regret' decision (Hänsch et al., 2000). The mediation report also examined alternatives to enlargement, including new runways to the south of the airport (which, ironically, would have meant the deconstruction of *Startbahn 18 West*). This southern alternative had become possible because of the withdrawal of US forces from Frankfurt in favour of their air-base in Ramstein in the 1990s. On the other hand, alternatives on the southern side of the airport had been considered as early as the 1960s and rejected as they would have required the relocation of a small town east of the airport. Nevertheless, the mediation board gave an indirect recommendation not to reject such options entirely out of hand (*ibid.*, p. 181).

As stated above, the night flight ban that came out of the mediation and which had been further examined in the RDF, was abandoned in the final plan approved in 2007. Later, in 2009, the administrative court considered that the mediation outcome should have been considered binding for the plan approval (Hessischer Verwaltungsgerichtshof, 2009a). Since the new protest movement focused on the night flight issue in particular, a closer look at it is necessary. The RDF issued a call for assessments of the practical feasibility and economic consequences of the plan in 2005. The resulting report showed several operational obstacles for the airlines but overall stated that a night flight ban was feasible (Regionales Dialogforum, 2004). Thereafter, the RDF developed a statement with regard to night flights that the mediation issue would mean to have no noise events between 11pm and 5am local time, and further that no take offs or landings would be permitted between 10:45pm and 4:44am local time (Regionales Dialogforum, 2005). Lufthansa refused to undersign the RDF statement (*ibid.*, p. 3). Following this, a judicial review was issued examining the implications of a night flight ban on the region. The experts concluded that it was lawful and inevitable especially if this would make the realisation of the northwest alternative for a new runway feasible (Hermes, 2006).[1]

FROM NIGHT FLIGHT BAN TO THE MAGIC 17

In December 2007, the Government of Hesse approved the final plan for the new northwestern runway: with the exception that 17 night flights would be permitted. It was argued that without these exceptions the nature of the hub airport itself was in danger as international cargo flows would be abandoned (Hessisches Ministerium für Wirtschaft, Verkehr und Landesentwicklung, 2007b, p. 24). Furthermore it was argued that the night flight ban was not 'court-proof' (*ibid.*, p. 28). It is worth mentioning that with this decision, the government did not follow the proposal that had been submitted by the airport operator, Fraport AG, who had requested a complete night flight ban.

With the admission of 17 exceptions from the night flight ban it could be said that the government achieved two things: it destroyed any trust in official institutions and reignited a debate that had already seemed tamed. The decision was immediately challenged at the Hessian administrative court in Kassel and then at the national administrative court in Leipzig.

After the opening of the new runway in October 2011 a new protest wave has developed. The opposition against the airport has reached higher social ranks and has, for example, caused high-ranking managers, all frequent flyers, to break with the traditional CDU during the election for the new mayor of Frankfurt. The story's end – so far – can be told quickly: the 'court-proof' exception turned out not to be court-proof – as well as the Hessian administrative court, but even beyond that, the national administrative court in Leipzig estimated the 17 exceptions as an *Abwägungsfehler*, a failure to balance competing demands – a sure slam on the brakes of any plan approval, and suggested that the government of Hesse revise the approval by holding another hearing with regard to the night flight ban (Bundesverwaltungsgericht, 2012). Table 1 summarises the main issues and turning points of airport development from 1936 to the present day.

RECENT PROTEST DEVELOPMENT

Practically with the opening of the new runway in October 2011, a completely new protest movement arose. New because the protagonists are now also coming from the wealthier towns along the Main-river-axis who have before been almost unaffected by aircraft noise. Also parts of Frankfurt, especially upward neighbourhoods in the southern part of the city, called 'Sachsenhausen', suddenly experienced a level of aircraft noise they have never expected. These areas, traditionally CDU-oriented, have seen a share in the change of power during the city mayoral election described in the introduction. A typical example is that of a retired couple, where the husband worked in international business – of course with air-travel-,

Table 1. Timetable Development of Frankfurt airport: Critical Turning Points and Incidents.

Date	Incident
1936	Establishment of the airport at today's location.
1965	Initiation of first enlargement procedure.
1984	Inauguration of the runway 18 West 'Startbahn West'.
17 Novemeber 1997	DER SPIEGEL-Interview with Lufthansa-chairman: Proposition for a new runway.
23 February 1998	First meeting of the expert council 'Gesprächskreis Flughafen'.
16 July 1998	Beginning of the mediation procedure.
January 2000	Final meeting of the mediation group and mediation report.
June 2000	Flotation of the airport management: From FAG to Fraport AG.
23 June 2000	Constitutional meeting of the regional dialogue-forum RDF.
22 October 2001	Beginning of regional planning procedure 'Raumordnungsverfahren'.
10 June 2002	End of regional planning procedure with preferation of the north-west runway.
10 September 2003	Beginning of airport plan approval procedure 'Planfeststellungsverfahren' applied by the Fraport AG with a complete night flight ban.
18 December 2007	Plan approval 'Planfeststellungsbeschluß': Night flight ban with 17 exemptions .
15 January 2009	Hessian administrative court refuses rush orders to stop the runway construction but expresses considerable concern about the approval of night flights (Hessischer Verwaltungsgerichtshof, 2009b).
20 January 2009–21 October 2011	Construction of the new runway north west, inauguration.
04 April 2012	National administrative court confirms the verdict in the sense that scheduled flights between 11 pm and 5 am are furthermore illegitimate (Bundesverwaltungsgericht, 2012).

who have for the first time in their lives elected the SPD. They own several houses in Sachsenhausen, including their own, and are elected as representative claimants by the administrative court. On the other hand, cities who have long been affected by airport noise, such as Rüsselsheim or Raunheim which have less wealthy inhabitants accuse the new protesters of having elected exactly the politicians who have always favoured the airport expansion.

Still the protest is not restricted to these new groups. Since the opening of the runway, almost every Monday protesters have demonstrated in the passenger terminal. The association of the citizen's initiatives are claiming

Fig. 5. A Carnival Float in the Shape of an Aircraft at a Parade in the Small Village of Oberhöchstadt Protests Against the Noise Created by 'Fraport Krachlines' (Krach = Colloquial for Noise). (Photograph : Author's own.)

in their website the 65th demonstration since the opening with around 1000 protesters.[2] Fig. 5 shows a critical carnival float at a village parade in 2012.

ANALYSIS

Within the legal framework of airport governance, the role of actors is crucial to tackle the question of possibilities and limitations of the system. Airport expansion and the accompanying conflicts over time are important to understand and evaluate the governance of airports. Since the development of FRA must be understood as a constant negotiation between changing coalitions and narratives play an important role here – narratives as a core part of spatially related conflicts (Reuber, 1999). Therefore, it is useful to have a look at how the constellations of actors change and rearrange.

Figs. 6–8 show which actors were active during different periods of the conflict and how they were bound together by a more or less stable institutional context. In 1980, the process of airport enlargement was mainly a task of the Hessian ministry of economy and traffic, which is tightly bound to the government of the Land Hessen itself. The airport company, at that

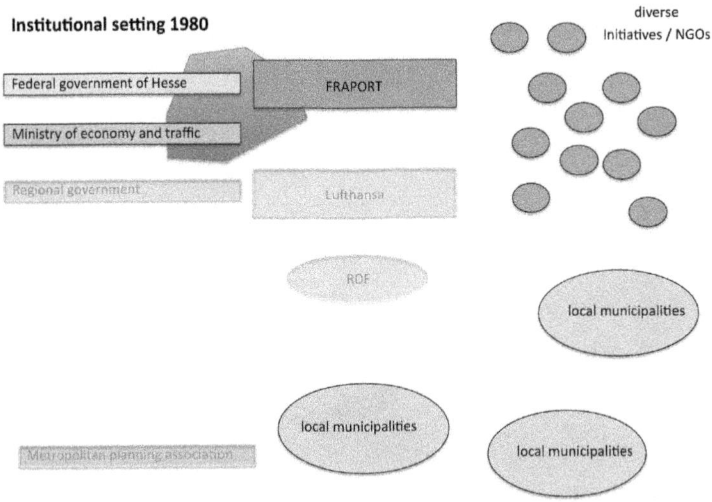

Fig. 6. Institutional Setting 1980s.

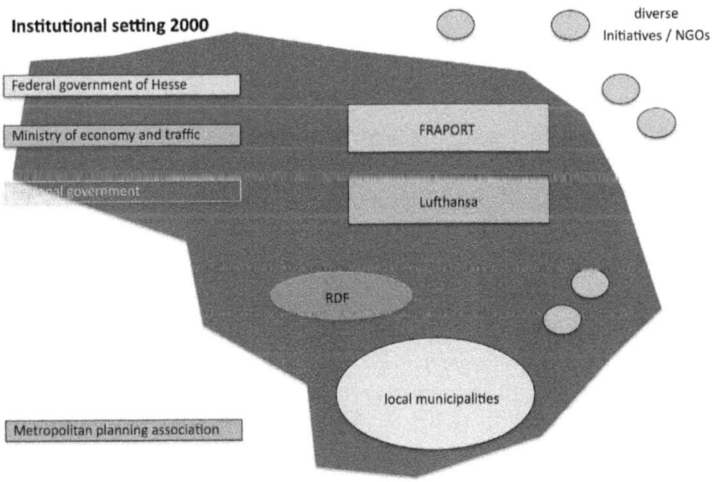

Fig. 7. Institutional Setting 2000.

time still 100% publically owned, was, of course, an important part but did not appear as an actor – merely the airport's interests were understood as a political will. A strong diversity of initiatives existed, a camp was built in the forest, meaning that some form of self-coordination happened.

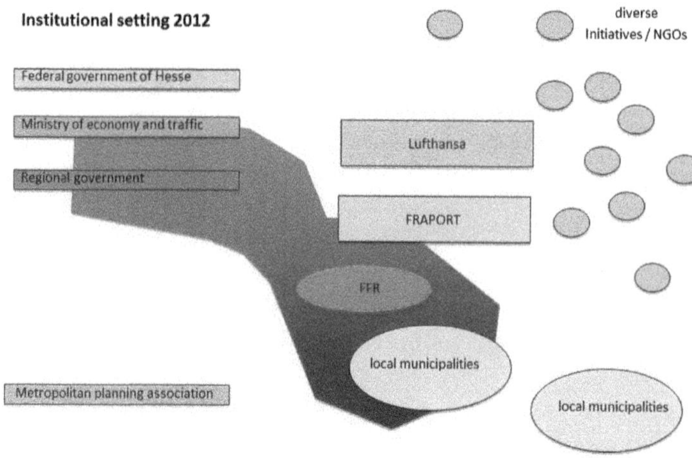

Fig. 8. Institutional Setting 2012.

The local municipalities were not in the centre of political decision making, but were rather marginal (Fig. 6). Lufthansa is not included since it did not appear as a strong actor, though it is inconceivable that it did not influence the process.

The setting in 2000 (Fig. 7) shows differences: the mediation procedure has been accomplished and the actors within the polygonal background area are part of the RDF. The number of initiatives has decreased as some merged into a joint civil initiative. Only few of them remained part of the mediation procedure, since they were unsatisfied with the procedure of member selection. This is understood as an inherent difficulty of the mediation. The selection of actors had in the first place been done by the minister of Hesse at that time, Hans Eichel of the SPD (Busch, 2000). The above mentioned expert round *Gesprächskreis Flughafen* had been selected by him rather individually than systematically, also because it was meant as a rather informal process in the first place. Even if here the RDF plays the institutionalising part of the structure, its role in the plan approval is not defined. The literally binding character of a dialogue as a social process was to be stated by the administrative courts (see above).

Lufthansa now appears to be an important actor, almost as relevant as the airport company itself. Lufthansa proposed the idea of a new runway and, as we will see, Lufthansa plays a central part in making an exception of the night flight ban. Also newly appearing is the Metropolitan Planning Association, the institution issuing the regional land-use plan (compare Fig. 3). This sort of plan as well as the institution are singular on this level in Germany. Usually the municipalities are in charge for land-use planning. This agency is issuing land-use plan on a metropolitan level and acts in some part in concurrence to the regional government, which is a substructure of the federal level.

Today the setting again has changed (Fig. 8): The former RDF has been changed to the *Forum Flughafen Region* (FFR), as a follow-up institution to the mediation and the RDF. Many municipalities left the forum after the plan approval, the nearby city of *Flörsheim* for example with their mayor stating: 'The FFR is a fig leaf for the ruthless prevailing of the air-transport industry's interests' (Rundschau, 2009).[3]

Also the participation in the RDF has not prevented actors from going to court after the plan was approved. It was Lufthansa especially who appears unwilling to mediate. One of their managers even stated in public that they would now rather exchange their night flights for a new runway. The company tried to delay the opening of the new runway, since it had not planned with the possibility of a realisation of the night flight ban (FAZ, 2011).

The question remains as to why the government of Hesse has taken the risk of reigniting a conflict for the sake of 17 cargo night flights. It is especially interesting since the government here has taken back the role it had in the 1980s, and has refrained from any attempt to stay in a neutral position like it did when the mediation started in the late 1990s. When state minister Roland Koch speaks in the dialogue forum's report *Waffengleichheit*, an equality of arms that (in his words) the mediation procedure has produced between the airport's opponents and the airport operator it can be argued that he may have misunderstood the idea of the procedure (Regionales Dialogforum, 2010). He must have returned to Geis (2003), who states in her analysis that the mediation, even though it seemed a failure to many actors, has decreased the informational asymmetry between the players. But this must be understood as a formerly unintended consequence or at least a side-effect. It is also doubtful if it is truly positive that opponents now are fighting about methodological relevances of expertise. The overall level of the debate has become more fact-oriented, but at the end it has not solved the friction.

NARRATIVES TOLD IN THE COURSE OF THE CONFLICT

The course of the mediation and the following procedures has destroyed trust within the region. It seems theoretical to ask if the furore against the new runway would have been less if the night flight ban would have been part of the approval. Suau-Sanchez (2010) presented findings that the perception of aircraft noise is influenced by the sense of trust people have in a political process. And states, with reference to Broer, that 'If policy treats a situation as problematic, people will experience it as problematic' (p. 139). Sack (2010) examines discourse formation in conflicts, pointing out two narratives that accompany the situation at Frankfurt airport: the airport as a global hub, that brings employment and transnational mobility, in danger by global competition, and the story of the region as a liveable neighbourhood, threatened by losing quality of life and depreciation. The mediation thus was the attempt to reach a 'win-win discourse' (see Sack, 2000). Whether it is possible or not to talk of an airport as being less noisy, with regard to the special character of spatial conflicts (see above). it is interesting that the discourse formations are structured by an image of the space.

It could be argued that the narratives created by both sides are developing a live of their own. If we look at the explanation given by government to allow an exception for 17 night flights, this narrative has been told by Lufthansa in the first place: 'The number of 17 flight movements between 11 pm and 5 am is indispensable and exigent to assure the connectivity of the airport to international cargo traffic through the here located home-base-carriers.' Giving two reasons: 'the strong linkages between these carriers and Frankfurt airport and because of investments already made at this location' (Hessisches Ministerium für Wirtschaft, Verkehr und Landesentwicklung, 2007b). Today Lufthansa is constructing a new cargo centre at the airport, a project that was only delayed a short time. In approving the proposal it was stated that: 'The interests of the applicant are congruent with the public interest' (Hessisches Ministerium für Wirtschaft, Verkehr und Landesentwicklung, 2007a).

Not appearing as a singular actor in the figures, but also important to the story is the City of Frankfurt, who strongly benefits from the airport, as well fiscally as a business location, without being too much affected by aircraft noise. Both the Land Hessen and the city of Frankfurt still own more than 50% of the shareholder value of Fraport. As a rule the mayor of Frankfurt is member of the supervisory board of Fraport.

The fear of losing the region's or city's position in global rankings and the association with the airport as an elixir against this danger can also be seen in local politics. Petra Roth, the Mayor of Frankfurt until early 2012, describes in a political meeting what she sees in the airport city: 'Airport cities will be a meeting point for innovators from all over the world (...).

Airport cities that have congress facilities, technology centres, incubators and exchange knowledge through fast transfer from continent to continent, are of course an important part of the value chain for cities' (PARLIS, 2008).

Here the description of the spatial image of an airport city has departed from the image of conventional airports and has become part of a narrative that could be called 'the global city concurrence threatening old Europe.' The meaningful interpretation of the airport city by the former mayor Petra Roth shows the extent to which the rhetorics of locational success and city concurrence have influenced the view of how local city politics should act.

These narratives are, of course, not exclusive; the airport opposition is also operating with strong notions of loss and gain. Sack (2010) detects biological analogies like terming the airport a 'malign ulcer'. With the opening of the new runway, a mayor of one nearby city draw a horror-vision of his city becoming like 'Raunheim', a settlement not very far away, that has suffered from airport noise for a long time. Unsurprisingly, the mayor of Raunheim was not amused.

Overall four main turning points in the development of the conflict and the spatial image of FRA can be detected. Firstly, the protest movement of the 1980s was focused on the destruction of the forest and aircraft emissions, but it was also about general values such as preservation of nature and local democracy and was placed in a framework of growing environmental care. During the 1990s the promise that the airport would grow within its existing boundaries was echoed, though the airport constructed the cargo city south, constraining physical development of the future airport. After the expansion proposal of 1997, the discussion tried to frame capacity enlargement as a 'win-win' situation, with the mediation attempting to negotiate the non-negotiable, developing a compromise between prosperity and growth on the one hand and the possibility for a liveable environment on the other. With the breakup of the package after the plan was approved in 2007, the protest has been atomised, resulting in self-organised rallies protesting even within the airport. The motive of protest is focused on the negative impact of aircraft noise disturbing one's daily life at home or in school. Concerns about the environmental impact of the airport are represented by NGOs such as the BUND (Bund für Umwelt und Naturschutz Deutschland) (German alliance for environment and natural preserve). The professionalsation of opposition during the mediation procedure has resulted in a differentiation of organisations with cognisance for all possible sub-interests.

STABILITY AND CHANGE IN THE REGIONAL GOVERNANCE CONTEXT

The decision making structure for airport governance has been relatively stable within recent years: no matter what party, the governments of Hesse have usually formed a growth-oriented coalition with the airport company. Together with the City of Frankfurt, a stable triangle has been formed ensuring the extension obverses the regional planning agency, opposition from civil activists and other municipalities. The formal structures of the sectorial procedure described above are supporting this growth coalition. This justifies the understanding of governance as the sum of different forms of regulation that are not isolated from each other: the character of sectorial procedures enables the actors in the regional context to enact what they perceive is the best for the economic prosperity of the region, giving preference to a decision that will almost inevitably produce protest among some parts of the society. The formally given power is executed not only because it is there, but because the decision is understood as painful for some individuals but overall believed to be necessary for the public interest. The public interest is expressed in the image of a region that is closely related to global value chains and a meeting point for innovators from all over the world.

The stability of this triangle has been profoundly shaken since the opening of the new runway: firstly, mayors of CDU-led cities in the areas surrounding the airport have started supporting the claims for a night flight ban. This has moved the government coalition to turn their position 180 degrees and also support the ban. But the relevance of the airport issue was completely underestimated by Petra Roth, who resigned surprisingly a year earlier than necessary as Mayor of Frankfurt, making way for the supposed to be crown-prince, the young minister of Hesse for internal affairs, Boris Rhein. But the candidate was perceived as part of the plan approval and the winning candidate, a former no-name parliamentarian of the SPD, Peter Feldmann, won the election with classical social democratic topics such as housing politics and child poverty, but also with the claim to extend the ban from 10pm to 6am for an hour, from 10pm to 6pm local time – a claim that his party, the SPD of Hesse, described as being unrealistic.

Even if the region has had a long experience with the character of spatial conflicts and infrastructure, the policy learning seems to have been undermined by a reliance on strong commitments. Now, once again, the airport has started changing political balances. The upcoming elections for the Hessian parliament in 2013 will show how much. Still, even opposition actors are not expecting airport governance in Rhine-Main to fundamentally change. Indeed, Franz Urhahn, the municipal councillor for Mörfelden-Walldorf (*Die Grünen*), states that once his party comes to power, the pressure from the airport industry will grow and politicians will rather buckle instead of being accused of jeopardising the economical position of Frankfurt Rhine-Main.[4] In brief, all Hessian state governments have, some more and some less, accounted the role of the airport as an important economical motor. And it is doubtful that any in the near future will. Obviously, the institution state government here structures the political agenda stronger than individual party's preferences.

IS AIRPORT GOVERNANCE POSSIBLE?

Looking at the conflict history of FRA and the impossibility of preventing the recent conflict by mediation, one could state that airport expansion conflicts are not at all amenable to mediation. The elaborated process of plan approval connected with numerous expertise about the future impact of the airport could not predict the effect of the real infrastructure and the sound of an airplane passing over somebody's roof. The idea of interest balancing that is inherent in the German planning systems also seems to come to an end: that is because the consequences for a house-owner feeling tortured by noise and deciding to sell the house and move away of a beloved environment are strong for the individual but hard to be weighed against an abstract employment multiplier. A governance structure is only as good as it is able to structure the decisions of the actor's involved. Two concluding comments can be made about the procedural framing and its application in the real case.

Firstly, time and alternatives: Planning large infrastructure projects, whether it is an airport or a train station, is a complex task. It is always hard to identify the right moment to go public with a plan or set of proposals. Almost irrespective, the 'usual protesters' will be there immediately. The broader public will, to a large extent, perhaps only notice when the excavator shovel hits the ground. Werner Rothengatter stated in a workshop on conflicts and infrastructure with regard to the case of the protest against the restructuring of the Stuttgart main-station that alternatives are not really discussed and that decisions are often taken under political rather than technical aspects.[5] This is a very crucial aspect: if it is possible to shift the public discussion from 'yes or no' to 'how'. In the case of FRA expansion one can state that the different alternatives have been biased at the moment they entered the discussion, which is another larger topic to discuss, that would exceed the scope of this chapter.

Secondly, clearing the role of informal processes and keeping on track: As stated above, a problem of the mediation was that it was not legally anchored in the formal procedure. How would the protest against the new runway have looked like if the government had not made the 17 exceptions? This is, of course, hypothetical, but the difference could have been huge. Now, the government has lost two very important lawsuits, with one court stating that even an informal mediation has to be accounted for in a formal procedure. Today the government faces a power change after elections in September 2013. The core outcome of a four-year process has been ignored – a time where many protagonists have spent many hours trying to mediate the 'unmediatable'.

When state minister Roland Koch stated in the dialogue forum's report of 2010 *Waffengleichheit*, an equality of arms that (in his words) the mediation procedure has produced between the airport's opponents and the airport operator it can be argued that he may have misunderstood the idea of the procedure (Regionales Dialogforum, 2010). He must have returned to Geis (2003), who states in her analysis that the mediation, even though it seemed a failure to many actors, has decreased the informational asymmetry between the players. But this must be understood as a formerly unintended consequence or at least a side-effect. It is also doubtful if it is truly positive that opponents now are fighting about methodological relevance of expertise. The overall level of the debate has become more fact-oriented but in the end this has not solved the friction.

To conclude, the German system of infrastructural planning is differentiated and is rooted in the idea to mirror the complexity of the governance of airport expansion. Still it lacks an idea for anchoring informal procedures and does not support the discussion of alternatives in public. It is in general able to cope with the complexity of the topic, but as the same state level that co-owns the airport often executes it, interest conflicts are possible. Also the ministries of economy, which might give it a bias, mostly undertake the procedures. A separate bureaucratic agency could be a solution to this.

An important aspect that the formal procedures also cannot cope with is the question of monetary substitution of private losses induced by the infrastructure. As Bickenbach et al. (2005) put it, this is a mayor aspect of reforming the German system. However all this might not reduce complexity, and the saying that while discussing about one airport in Germany, 10 are built in China remains a simplistic but ubiquitous opinion. In the light of projects as FRA, the massive protest against the Stuttgart 21 main-station and not to forget the upcoming number of infrastructures necessary for the *Energiewende,* the demand for public participation has reached a new boom. At the same time this is raising questions how those projects in the future can be executed at all in a society that is characterised by economic maturity and demographic shrinkage.[6]

NOTES

1. After the mediation report, it soon became clear that of all the examined alternatives the north-western runway was the government's preferred option. One possible theory is that alternatives in the south would have forced Fraport to give up their very lucrative cargo real-estate in the south of the airport (Knippenberger, 2012).
2. http://www.flughafen-bi.de/ association of civil initiatives with many pictures of the so-called 'Montagsdemos'. The term is misleading and unfit, since it was the Monday demonstrations 'Montagsdemo' of civilians in the former GDR, who started the end of the regime and the socialist dictatorship.
3. Own translation
4. Personal conversation with Erster Stadtrat Franz Urhahn 23 November 2012.
5. Rothengatter is emeritus of the University of Karlsruhe, today Karlsruhe Institute of Technology KIT. He took part in a workshop organised by the author at KIT in December 2010, presenting the case 'Stuttgart 21', on which he was chosen as an expert by the national ministry of Transport BMVBS, retrieved from http://www.iww.uni-karlsruhe.de/mitarbeiter_114.php, last visited 18 June 2013.
6. For a good status quo compare: *The Economist* (2013): The reluctant hegemon: a special report on Germany.

REFERENCES

Bickenbach, F., Kumkar, L., Sichelschmidt, H., Soltwedel, R., & Wolf, H. (2005). *Ausbau der Flughafeninfrastruktur*. Berlin: Springer.

BMVBS, B. für V., Bau und Stadtentwicklung. (2009). Flughafenkonzept der Bundesregierung 2009. Retrieved from http://www.bmvbs.de/Anlage/original_1090412/Flughafenkonzept-2009-der-Bundesregierung.pdf

Bundesverwaltungsgericht. (2012, April). Pressemitteilung Nr. 33/2012. Retrieved from http://www.bverwg.de/enid/3f32ac7481ed9f20b32bdec2c7bca678,cb072c7365617263685f646973706c6179436f6e7461696e6572092d09313431353909 3a095f7472636964092d09353737/Pressemitteilungen/Pressemitteilung_9d.html

Busch, P.-O. (2000). *Konfliktfall Flughafenerweiterung eine kritische Würdigung des Verfahrens Mediation - eine Zukunftsregion im offenen Dialog zum Flughafen Frankfurt, Main.* Frankfurt am Main: Hess. Stiftung Friedens- und Konfliktforschung.

Button, K. (2002). Debunking some common myths about airport hubs. *Journal of Air Transport Management*, *8*(3), 177–188. Retrieved from http://www.sciencedirect.com/science/article/B6VGP-44XC8XF-1/2/69d9f49969fd7cdf4e4b0af0a08ab521.

Button, K., & Lall, S. (1999). The economics of being an airport hub city. *Research in Transportation Economics*, *5*, 75–105. doi:10.1016/S0739-8859(99)80005-5.

Der Spiegel. (1997). Neue Startbahn. *Der Spiegel*, *47*, 126–128. Retrieved from http://www.spiegel.de/spiegel/print/d-8823145.html

Einig, K., Grabher, G., Ibert, O., & Strubelt, W. (2005). Urban governance. *Informationen zur Raumentwicklung*, *9/10*, S576–S577.

FAZ. (2011, 17 October). Lufthansa und Fraport geraten aneinander: Schatten über neuer Landebahn in Frankfurt. *Frankfurter Allgemeine Zeitung*. Retrieved from http://www.faz.net/aktuell/wirtschaft/lufthansa-und-fraport-geraten-aneinander-schatten-ueber-neuer-landebahn-in-frankfurt-11496632.html

Fraport AG. (2013). Zahlen, Daten, Fakten 2013 zum Flughafen Frankfurt. *Fraport AG*. Retrieved from http://www.fraport.de/zahlendatenfakten/. Accessed on 16 June 2013.

Fürst, D. (2007). Urban Governance – Einblicke in die Diskussion. Magazin Städte im Umbruch. *Urban Governance und Steuerungsformen (in) der Stadtentwicklung*, *4*, 5–10.

Geis, A. (2003). *Umstritten, aber wirkungsvollé: die Frankfurter Flughafen-Mediation*. Frankfurt: Hessische Stiftung Friedens- und Konfliktforschung.

Hänsch, K., Niethammer, F., & Oeser, K. (Eds.). (2000). Bericht Mediation Flughafen Frankfurt/Main. Mediationsgruppe, Flughafen Frankfurt/Main. Retrieved from http://www.forum-flughafen-region.de/service/archive-mediation-rdf/das-mediationsverfahren-flughafen-frankurt/

Hermes, G. (2006). Rechtsfragen der Verankerung verbindlicher Ziele im Landesentwicklungsplan Hessen: Rechtsgutachten erstattet im Auftrag des Regionalen Dialogforums Flughafen Frankfurt. Retrieved from http://www.forum-flughafen-region.de/service/archiv-des-regionalen-dialogforums/gutachten-und-positionen-des-regionalen-dialogforums/

Hessischer Rundfunk. (n.d.). 75 Jahre Flughafen Frankfurt: 1936–2011 | hr. *hr-online*. Retrieved from http://www.hr-online.de/website/specials/75-jahre-flughafen/index.jsp. Accessed on 11 November 2012.

Hessischer Verwaltungsgerichtshof. Bau der neuen Landebahn steht nichts entgegen Voraussichtlich aber keine planmäßigen Flüge in der Kernnacht. Pressemitteilung zum Beschluss B 254 08 T vom 15. Januar 2009. , § Nr.: 3/2009 (2009a). Retrieved from http://bbi.unser-forum.de/Archiv/2009/090115_1543_flughafen_urteil.pdf

Hessischer Verwaltungsgerichtshof. Beschluss vom 15.01.2009 11 B 254 08 T., § Nr.: 3/2009 (2009b). Retrieved from http://bbi.unser-forum.de/Archiv/2009/090115_1543_flughafen_urteil.pdf

Hessisches Ministerium für Wirtschaft, Verkehr und Landesentwicklung. Planfeststellungsbeschluss für den Ausbau des Verkehrsflughafen Frankfurt Main., Pub. L. No. PF-66 p-V (2007a). Retrieved from http://www.wirtschaft.hessen.de/irj/HMWVL_Internet?uid=5b625deb-9ee6-11d5-ce7b-91921321b2c3

Hessisches Ministerium für Wirtschaft, Verkehr und Landesentwicklung. (2007b, December 18). Sprechzettel Hessens Wirtschaftsminister Dr. Alois Rhiel Pressegespräch Entscheidung über den Planfeststellungsantrag zur Erweiterung des Flughafens Frankfurt Main; Landeshaus, Wiesbaden. Retrieved from http://www.wirtschaft.hessen.de/irj/HMWVL_Internet?uid=5b625deb-9ee6-11d5-ce7b-91921321b2c3

Hughes, T. (1987). The Evolution of Large Technological Systems. In W. E. Bijker, T. P. Hughes, & T. J. Pinch (Eds.), *The social construction of technological systems: New directions in the sociology and history of technology*. Cambridge, MA: MIT Press. Retrieved from http://arno.unimaas.nl/show.cgi?did = 3599

Johnsen, H. (1996). *Der Startbahn-West Konflikt: ein politisches Lehrstück?: Zeitzeugen ziehen Zwischenbilanz*. Frankfurt am Main: Societäts-Verlag.

Knippenberger, U. (2012). *Regionale Governance des Funktionswandels von Flughäfen: Eine Analyse am Beispiel der "Airport City" Frankfurt am Main*. Wiesbaden: VS Verlag fuér Sozialwissenschaften.

Langhagen-Rohrbach, C. (2002). Entwicklung der Sozial- und Wirtschaftsstruktur in den umliegenden Gemeinden des Flughafens Frankfurt. KSR36-Heft 5 Institut für Kulturgeographie, Stadt- und Regionalforschung Johann Wolfgang Goethe-Universität. Retrieved from http://www.geo.uni-frankfurt.de/ifh/Forschung/Publikationen/Archiv/KSR36/index.html

Mayntz, R. (1993). Grosse technische Systeme und ihre gesellschaftstheoretische Bedeutung. *Kölner Zeitschrift für Soziologie und Sozialpsychologie, Jg. 45*(H. 1), S. 97–108.

Mayntz, R. (2004). *Governance Theory als fortentwickelte Steuerungstheorie?* Retrieved from http://www.mpifg.de/pu/workpap/wp04-1/wp04-1.html

Mayntz, R., & Scharpf, F. (1995). Der Ansatz des akteurzentrierten Institutionalismus. In *Gesellschaftliche Selbstregelung und politische Steuerung*. Frankfurt am Main: Campus.

Neff, A. (1997). *Zwangsarbeit am Flughafen Rhein-Main im Zweiten Weltkrieg*. Archiv für hessische Geschichte und Altertumskunde, N.F. 55, 111–147. Retrieved from http://publikationen.ub.uni-frankfurt.de/frontdoor/index/index/docId/16591

PARLIS, P. (2008). *Wortprotokoll zur 19*. Sitzung der STVV am 16.01.2008. Retrieved from http://www.stvv.frankfurt.de/parlis/parlis.htm

Regionales Dialogforum. (2004). Praxisorientiertes Umsetzungskonzept zur Verlagerung der Flugbewegungen in den Zeiten des geplanten Nachtflugverbotes am Flughafen Frankfurt, im Auftrag der IFOK GmbH im Rahmen des Regionalen Dialogforums. Retrieved from http://www.forum-flughafen-region.de/service/archiv-des-regionalen-dialogforums/gutachten-und-positionen-des-regionalen-dialogforums/

Regionales Dialogforum. (2005, September). RDF Papier zur Bewertung der Ergebnisse des Gutachtens "Praxisorientiertes Umsetzungskonzept zur Verlagerung der Flugbewegungen in den Zeiten des geplanten Nachtflugverbots" und zum weiteren Vorgehen. Verabschiedet 23. Retrieved from http://www.forum-flughafen-region.de/service/archiv des regionalen dialogforums/gutachten-und-positionen-des-regionalen-dialogforums/

Regionales Dialogforum. (2010). *Abschlussdokumentation RDF 2000–2008*. Retrieved from http://www.forum-flughafen-region.de/service/archiv-des-regionalen-dialogforums/rdf/

Reuber, P. (1999). *Raumbezogene politische Konflikte: Geographische Konfliktforschung am Beispiel von Gemeindegebietsreformen*. Stuttgart: Steiner.

Rundschau, F. (2009, December 22). Flughafen-Forum: Flörsheim steigt aus. *Frankfurter Rundschau*.

Sack, D. (2000). *Startbahn West + 15 Jahre + Mediation = "Win-win-Paket"?* Kommune. Forum für Politik, Ökonomie, Kultur.Retrieved from http://www.oeko-net.de/kommune/

Sack, D. (2002). *Glokalisierung und Konfliktvermittlung*. In "Empirische und normative Analysen politischer Konflikte" des Arbeitskreises "Handlungs- und Entscheidungstheorien" der DVPW am 7./8.6.2002 an der Universität Mainz. Mainz: Universität Mainz. Retrieved from http://www.politik.uni-mainz.de/Bereich/bereich1213.bak/ak/Sack%20Manuskript.htm

Sack, D. (2010). Discourse formation in spatial conflicts. Presented at the Mobility infrastructures and spatial conflicts, Workshop at KIT Karlsruhe, Karlsruhe.

Scharpf, F. W. (2006). *Interaktionsformen: Akteurzentrierter Institutionalismus in der Politikforschung*. Wiesbaden: VS Verlag für Sozialwissenschaften.

Suau-Sanchez, P. (2010). Subjective appraisal of aircraft noise: From decibel measurements to effective noise management. In U. Knippenberger & A. Wall (Eds.), *Airports in Cities and Regions – Research and Practise* (pp. 137–142). Karlsruhe: KIT-Verlag.

Wolf, H. (2003). *Privatisierung im Flughafensektor: eine ordnungspolitische Analyse*. Berlin: Springer.

Ziekow, J. (Ed.). (2002). *Flughafenplanung, Planfeststellungsverfahren, Anforderungen an die Planungsentscheidung Vorträge auf den Dritten Speyerer Planungsrechtstagen und dem Speyerer Luftverkehrsrechtstag vom 21. bis 23. März 2001 an der Deutschen Hochschule für Verwaltungswissenschaften Speyer*. Berlin: Duncker & Humblot.

ABOUT THE AUTHORS

Peter Adey is Professor of Human Geography in the Politics, Development and Sustainability Group in the Geography Department at Royal Holloway, University of London. His work lies at the intersection between space, security and mobility and the blurring boundaries between cultural and political geography. Much of his research has contributed to the putative 'new mobilities paradigm', particularly as it relates to the sites, spaces and experiences of air travel, and he has published extensively on the subject. His current research focuses on three main empirical and conceptual sites of enquiry: the airport/border; the vertical territorial and material spaces of air; and the political and technological spaces of emergency and evacuation. In 2011 he was awarded a Philip Leverhulme Prize.

Kevin Anderson is Professor of Energy and Climate Change in the School of Mechanical, Aerospace and Civil Engineering at the University of Manchester. He is a deputy director of the Tyndall Centre for Climate Change Research and is research active with recent publications in Royal Society journals and Nature. He engages widely across all tiers of government; from reporting on aviation-related emissions to the EU Parliament, advising the prime minister's office on carbon trading and having contributed to the development of the UK's Climate Change Act. With his colleague Alice Bows-Larkin, his work on carbon budgets has been pivotal in revealing the widening gulf between political rhetoric on climate change and the reality of rapidly escalating emissions. His work makes clear that there is now little chance of maintaining the rise in global temperature at below 2°C, despite repeated high-level statements to the contrary. Moreover, his research demonstrates how avoiding even a 4°C rise demands a radical reframing of both the climate change agenda and the economic characterisation of contemporary society. He has a decade's industrial experience, principally in the petrochemical industry. He sits as commissioner on the Welsh government's climate change commission and is a director of greenstone carbon management. He sits as commissioner on the Welsh government's climate change commission, is a director of greenstone carbon management and a fellow of the Institution of Mechanical Engineers.

Annela Anger-Kraavi is a Lecturer in Macroeconomics at the University of East Anglia and college research associate in economics, Emmanuel College, University of Cambridge, UK. Her research interests include applied macroeconomics; large scale energy-environment-economy modelling; modelling climate change policy impacts; climate change mitigation policies for international aviation and shipping; and the co-benefits of climate change mitigation policies such as impacts on atmospheric pollution. She teaches environmental economics and has worked on projects on climate change policies commissioned by the UN and UK government. She holds BSc and MA in biology, BSc in economics, MPhil in environmental policy and a PhD in applied macroeconomics from the University of Cambridge. Prior to joining the University of East Anglia in 2013 she was a senior research associate at the University of Cambridge.

John Bowen is an Associate Professor in the Department of Geography at Central Washington University, USA. He received his PhD from the University of Kentucky and has taught at several universities in the United States and Singapore. He also has airline industry experience, having worked in cargo marketing for Singapore Airlines in Singapore in the 1990s. Much of his research has concerned air transportation, with particular emphases on the role of aviation in economic development, airline industry liberalization in developing economies, and the changing geography of commercial aircraft production. In 2010, his book *The Economic Geography of Air Transportation: Space, Time, and the Freedom of the Sky* was published by Routledge.

Alice Bows-Larkin is a Reader in Energy and Climate Change as part of the Tyndall Centre for Climate Change Research and based within the School of Mechanical, Aerospace and Civil Engineering (MACE), University of Manchester, UK. She trained as an astrophysicist at the University of Leeds, did her PhD in climate modelling at Imperial College, then worked in science communication. She returned to academia in 2003 joining the interdisciplinary Tyndall Centre to research conflicts between climate change and aviation. In 2008 she was appointed as a lecturer directing projects on international shipping and food supply scenarios within a climate change context. She makes an explicit connection between her academic research and the broader policy context and, as a result, her work on international transport, energy systems and carbon budgets has had a significant impact on topical debates and policy development, including shaping the UK's Climate Change Act. She currently leads research on decarbonising international shipping as part of the EPSRC's Energy

Programme, heads up, MACE's infrastructure research theme as well as food@manchester and is an associate editor of the journal *Carbon Management*.

Christian Bröer is an Associate Professor of Sociology at the University of Amsterdam. He researches the emergence of problems with noise, (mental) health and risks as an interplay between institutions and everyday life concerns. His research has appeared in the *Journal of the Acoustical Society of America*, *Sociology of Health and Illness* and *Mobilization* among others. He is currently running research projects on: contestations about mobile phone technology (funded with €360,000 by the Dutch research council), risk policies, ADHD in a comparative perspective and the everyday life of sadness and graffiti. He is head of the research group Dynamics of Citizenship and Culture at the University of Amsterdam.

Lucy Budd is a Senior Lecturer in Air Transport in the Transport Studies Group in the School of Civil and Building Engineering at Loughborough University, UK. She has a long-standing interest in aviation and her PhD, obtained in 2007, explored the multiple and contested geographies of airspace. Since then, she has developed theoretical and empirical research into airspace and air transport operations. Much of her current research focuses on the operational practices and environmental implications of air travel at a variety of sites and scales. Her research has been published in over 25 peer-reviewed journals, academic books, official reports and consumer publications and she has been interviewed on national television and radio on the subject of flight.

Thomas Budd is a Research Fellow in the Centre for Transport Research at the University of Aberdeen, specialising in issues of sustainable aviation and travel behaviour. His work has been published in a range of academic and industry titles, including the *Journal of Airport Management* and *Research in Transportation Business and Management*, and has been presented at a number of international conferences in Europe and North America. His PhD, conducted in the Transport Studies Group at Loughborough University, examined passenger surface access travel behaviour at Manchester Airport and explored options for reducing private vehicle journeys.

James Connelly is Professor of Politics at the University of Hull where he teaches political theory and environmental politics. He has published widely in political theory, environmental politics and the philosophy of

R.G. Collingwood and other idealist philosophers. His recent publications include *Politics and the Environment: From Theory to Practice* (with G. Smith, D. Benson and C. Saunders, 3rd ed., Routledge, 2012); *The Withering of the Welfare State: Regression* (ed. with J. Hayward, Palgrave, 2012); and *The EU as a Leader in Climate Change Policy* (ed. with R. K. Wurzel, Routledge, 2011).

Frédéric Dobruszkes is a Post-Doctoral Marie Curie researcher at the Oxford University Transport Studies Unit, UK. He is also a part time lecturer at the Free University of Brussels (ULB) and vice-president of the Brussels Regional Commission of Mobility. He has an MA in geography (1995), an MA in transport (1996) and a PhD in geography (2007) from the Free University of Brussels. His main research interests relate to transport geography and policy. His current focus is on the dynamics of the European air transport sector and on airline/high-speed rail competition.

Moshe Givoni is a Senior Lecturer at the Department of Geography and the Human Environment and head of the Transport Research Unit at Tel-Aviv University in Israel. He is also a visiting research associate at the Transport Studies Unit, which is part of the School of Geography and the Environment (SoGE) at the University of Oxford, and an associate editor for *Transport Reviews* journal. Before joining Tel-Aviv University in October 2011, he was a Senior Research in Transport Policy and Economics at the Transport Studies Unit and a research fellow at Wolfson College, both at the University of Oxford. He joined the Transport Studies Unit in October 2007, after completing a two-year Marie Curie Intra European Fellowship at the Department of Spatial Economics, Free University Amsterdam. He gained his PhD at the Bartlett School of Planning, University College London in 2005.

Steven Griggs is a Reader in Local Governance in the Local Governance Research Unit at De Montfort University in Leicester. His research evaluates the contribution of political discourse theory to understandings of the policy process, analysing, in particular, how shared meanings or institutional rules and norms emerge to 'make sense', privilege or constrain different democratic spaces within the policy process. Empirically, he has investigated these theoretical questions through research into two main areas of study; the politics of sustainable aviation and the expansion of airports in the United Kingdom and the dynamics of participation in neighbourhood governance and the democratic limits of the emerging discourse of community cohesion and sustainable communities. He has published extensively in the area of airport community protest and he is co-editor of the journal *Critical Policy Studies*.

Charlotte Halpern holds a PhD in political science and is an FNSP Associate Research Professor at the Centre d'Etudes Européennes de Sciences Po in Paris. Her main research interest is state restructuring and policy change in Europe. She has conducted extensive research on various infrastructure policies (airports, waste disposal and light rail) and environmental policy instruments in several European member states and the EU. She has published on comparative European policies and EU governance in leading political science journals such as *West European Politics*, *Environment and Planning C* and the *Revue Française de Science Politique*. She is currently working on a single authored book on comparative airport policies in Europe, and co-edits a forthcoming special issue of *Comparative European Politics* on new modes of governance in Europe (2015). She teaches comparative public policies at Sciences Po Paris and Grenoble and is the scientific director of the Sciences Po Urban and Regional Management Master Programme.

David Howarth is a Reader in the Department of Government and co-director of the Centre for Theoretical Studies at the University of Essex. He is first and foremost a political theorist, whose main interests are poststructuralist theories of society and politics, focussing especially on the empirical study of political ideologies and discourses; the theoretical relationship between space, time and the political; and the intersections between identity, difference and subjectivity. He has written a single-authored book entitled *Discourse* (Buckingham, Open University Press, 2000), which puts forward a distinctive theory of discourse and an approach for its analysis, and he has recently co-authored (with Steven Griggs) a new book on the *Politics of Sustainable Aviation* (Manchester University Press, 2013).

Ute Knippenberger received her diploma in architecture at the Technical University of Darmstadt, Germany, in 1999. After several years of urban design and planning practice she took up a PhD course at the University of Weimar, receiving her degree as a doctor in engineering in 2011. Her doctoral thesis, published in 2012, examined the regional governance of airports using the example of Frankfurt/Main and made special reference to the changing character of airports from traffic nodes to multifunctional 'airport cities'. Between 2007 and 2011, she taught urban design and planning at the Karlsruhe Institute of Technology. Since 2011 she has returned to planning practice as head of the Department of Urban Planning and

Environment at the city of Kronberg im Taunus, a midsize town within the urban agglomeration of Frankfurt.

Jonathan Köhler is a Senior Scientist at the Fraunhofer ISI (Institute for Systems and Innovation research). He has a PhD in 'Bounded Rationality in Savings Decisions'. From 2000 to 2005 he was the research theme manager, integrating frameworks, at the Tyndall Centre, responsible for development and coordination of the research theme on integrated assessment methodologies for climate change policy analysis and management of 12 research projects within the theme. He has worked on Integrated Assessment Model (IAM) development for climate policy and on EU and global macroeconomic modelling for energy and climate policy analysis. He was theme leader of the economics theme in the UK OMEGA research consortium on aviation and the environment and he is now working on transitions modelling and the modelling of innovation systems and processes in transport. He is involved in the EU CleanSky research consortium and has published on emissions trading in aviation.

INDEX